SENTIENT CITY

UBIQUITOUS COMPUTING, ARCHITECTURE, AND THE FUTURE OF URBAN SPACE

D1326846

SENTIENT CITY: ubiquitous computing, architecture, and the future of urban space

Edited by Mark Shepard

Case Studies by
David Benjamin, Soo-in Yang, and Natalie Jeremijenko
Haque Design + Research
SENSEable City Lab
David Jimison and JooYoun Paek
Anthony Townsend, Antonina Simeti, Dana Spiegel, Laura Forlano, and Tony Bacigalupo

With contributions by
Martijn de Waal
Keller Easterling
Matthew Fuller
Anne Galloway
Dan Hill
Omar Khan
Saskia Sassen
Trebor Scholz
Haddas Steiner
Kazys Varnelis
Mimi Zeiger

This book expands on an exhibition organized by the Architectural League and curated by Mark Shepard that was presented in New York City in Fall 2009.

The exhibition was made possible with generous support from the J. Clawson Mills Fund of the Architectural League and the Graham Foundation for Advanced Studies in the Fine Arts.

Additional support was provided by the Department of Architecture, School of Architecture and Planning, and the Department of Media Study, College of Arts and Sciences at the University of Buffalo. Support was also provided by public funds from the New York City Department of Cultural Affairs, in partnership with the City Council.

Additional support for production of this book provided by the Julian Park Publication Fund and the Digital Humanities Initiative at Buffalo, College of Arts and Sciences, University at Buffalo

Copublished by
The Architectural League of New York
594 Broadway, Suite 607
New York, NY 10012
212 753 1722
www.archleague.org
and
The MIT Press
Massachusetts Institute of Technology
Cambridge, Massachusetts 02142
http://mitpress.mit.edu

©2011 The Architectural League of New York
Contributions ©2011 by respective authors

Library of Congress Cataloging-in-Publication Data

Sentient city : ubiquitous computing, architecture, and the future of urban space / edited by Mark Shepard.
p. cm.
Includes index.
ISBN 978-0-262-51586-3 (pbk. : alk. paper)
1. Architecture and technology. 2. Architecture and society. 3. Information technology—Social aspects. 4. Cities and towns—Effect of technological innovations on. I. Shepard, Mark. II. Title: Ubiquitous computing, architecture, and the future of urban space.

NA2543.T43S47 2011
720.1'05—dc22
2010039579

Design: Thumb
Printed and bound in Hong Kong

AMBLER LIBRARY

SENTIENT CITY

ubiquitous computing, architecture, and the future of urban space

(AMB)
NA
2543
T43
S47
2011

Edited by MARK SHEPARD

The Architectural League of New York

The MIT Press • Cambridge, Massachusetts • London, England

CONTENTS

CASE STUDIES

ESSAYS

ACKNOWLEDGMENTS

This book results from the gracious efforts of many people.

The Architectural League of New York has been an avid supporter from the start of the project, beginning with the Architecture and Situated Technologies symposium held at the Urban Center in 2006, continuing through the Situated Technologies Pamphlets Series that they publish, and of course its role as organizer of the 2009 exhibition that I curated at the Urban Center, Toward the Sentient City, from which much of this book emerged. In particular, without the sustained and unyielding support of Rosalie Genevro, Executive Director at the League, together with the patient and persistent efforts of Gregory Wessner, Digital Programs and Exhibitions Director, this book would not be in print.

I am extremely grateful to the MIT Press and Roger Conover for believing in this project and agreeing to co-publish this book with the League.

Special thanks are due to Luke Bulman and Jessica Young at Thumb, exhibition designers for Toward the Sentient City as well as designers of this book, who not only have been a pleasure to work with but have also, in their deep understanding of the context and aspirations of this project, given it a form that few others could.

My colleagues Omar Khan and Trebor Scholz, with whom I co-organized the Architecture and Situated Technologies symposium and co-edited the Situated Technologies Pamphlets Series, have been instrumental in helping to shape the questions that both the exhibition and this book have attempted to address, and I continue to be thankful for their generous contributions to our ongoing collaborations.

Thanks to the members of the Toward the Sentient City selection panel, who generously donated their time to review more than 150 proposals from around the world and to participate in a rigorous and lively discussion that resulted in the selection of the commissioned projects: Amanda McDonald Crowley, Keller Easterling, Rosalie Genevro, Omar Khan, Laura Kurgan, Trebor Scholz, and Gregory Wessner.

Much appreciation is due to the participants in the Architecture and Situated Technologies Symposium: Jonah-Brucker Cohen, Richard Coyne, Michael Fox, Karmen Franinovic, Anne Galloway, Charlie Gere, Usman Haque, Peter Hasdell, Natalie Jeremijenko, Sheila Kennedy, Eric Paulos, and Kazys Varnelis; to the authors of the Situated Technologies Pamphlets Series (to date): Adam Greenfield, Usman Haque, Matthew Fuller, Benjamin Bratton, Natalie Jeremijenko, Laura Forlano, Dharma Dailey, Philip Beesley, Jullian Bleecker, Nicolas Nova, Marc Böhlen, Hans Frei, and Laura Y. Liu; and to the Situated Technologies Pamphlets advisory committee: Keller Easterling, Anne Galloway, Malcolm McCullough, and Howard Rheingold.

Sincere thanks are due to the University at Buffalo, The State University of New York. Thanks to my departmental chairs, Mehrdad Hadighi (Architecture) and Roy Roussel (Media Study), who have given me the time to focus on the research and production of this book, and to Brian Carter, Dean of the School of Architecture and Planning, who provided thoughtful advice along the course of this book's preparation.

Adam Jacob Levin, my research assistant, provided invaluable administrative assistance, copy-editing and image research. His attention to detail has been invaluable in preparing this book for production and his thoughtful insights with respect to its contents have helped shaped its final form.

Finally, I would like to thank my family, friends and colleagues who in so many different ways have provided the kinds of support without which a project like this would not be possible.

— Mark Shepard

PREFACE

GREGORY WESSNER

Sentient City: Ubiquitous Computing, Architecture, and the Future of Urban Space expands on a Fall 2009 exhibition organized by the Architectural League and curated by Mark Shepard. Like the exhibition on which it is based, the book questions the ways in which ubiquitous computing is transforming the design and use of buildings and cities. As the second decade of the twenty-first century begins, it is abundantly clear that a diverse range of digital technologies are operating beyond what many of us long accepted as the traditional site of computing — the desktop or laptop PC. In the span of a decade it seems that everything in our lives is now "smart": our phones, our appliances, our running shoes, our cars. We are accustomed to finding advanced technology in places that would otherwise have been unlikely — if not unthinkable — just a couple of years ago.

As pervasive as digital technology may be, however, there is still confusion and uncertainty about what the long-term implications will be for the built environment. What does any of this have to do with architecture? For many, it is a leap of the imagination to think that a microprocessor the size of your fingertip, or the mobile phone in your pocket, can meaningfully affect the shape of the room you're sitting in, let alone a city's skyline. At a time when digital technologies seem to be dematerializing more and more of the world around us (think books, CDs, photographs), what impact can they possibly have on the inevitable materiality of buildings and cities?

In The Big Switch: Rewiring the World from Edison to Google, author Nicholas Carr suggests that the advent of the electrical grid in the nineteenth century offers parallels for understanding the possible implications of the technological revolution we are currently experiencing. Carr argues that while the nation's grid was originally developed primarily as a means to provide safe and accessible power to industry, it set off a chain reaction of unanticipated effects that ultimately transformed American society. In the realm of buildings and cities, these ranged from the proliferation of artificial lighting that opened up cities to new forms of outdoor nightlife, to the provision of power for the elevators that enabled buildings to grow taller and the machines that displaced craft workers from their roles in production and fabrication. And Carr contends that the electrical grid fostered even more profound and long-lasting changes. "The rise of the middle class, the expansion of public education, the flowering of mass culture, the movement of the population to the suburbs, the shift from an industrial to a service economy," Carr argues, "None of these would have happened without the cheap current generated by utilities." All of which, in turn, were reflected in transformations in the built environment.

We are now on the cusp of a similarly fundamental reconfiguration of physical space, one in which a vast and mostly invisible layer of technology is being embedded into the world around us. Using a wide range of complex technologies and devices — from microprocessors and electronic identification tags to sensors and networked information systems — buildings and cities are being transformed, imbued with the capacity to sense, record, process, transmit, and respond to information and activity taking place within and around them. What unanticipated effects will emerge this time around and how will they manifest themselves in buildings and cities? How will an increasingly mobile workforce, untethered from the office, alter the configuration of workspace and the design of office buildings? As online commerce expands, how will changes in the way we consume affect the size and layout of stores, as well as the distribution networks and facilities that supply them? How will technology infrastructures reorient cities and transform the way we navigate them? If we instrumentalize the physical world around us, animating furniture and rooms and building façades with the ability to collect data, process information, and take action — what profound changes will this bring about in the shape of our built environment and how we exist in it?

Of course, predictions for a future "smart" city have been floating around for decades, and we are all familiar with the false starts and wrong turns. What makes this moment unlike any before is that the decreasing cost of the hardware and the increasing computational power of the software have converged so that it is now feasible to embed enormously powerful digital intelligence and processing capability into seemingly any object or space. If experience has taught us anything, it is that new technologies get integrated into the existing built fabric in complex and unexpected ways. How will the existing built fabric be retrofitted to enable new technologies? What new forms will emerge in our future buildings? As the "sentient" city becomes a reality, the questions we need to ask now are: what will it look like, how will it work, and who will benefit from it?

Our hope is that this book will challenge all of us to begin thinking about alternative ways to answer these and other questions. We also hope that it will bring architects into a conversation that until now has been dominated by technologists. That these technologies will continue to permeate our world is inevitable. The possibilities for integrating our lives into a networked whole, for increasing safety and security, for off-setting environmental degradation, are too great to forego — and the potential for economic gain will be too seductive to resist. We should not let the technology (or the terminology) mislead us into thinking that these are issues relevant, and accessible, only to the technorati. What we are talking about is nothing short of a complete reorientation of our relationship to the built environment. Architects and urban designers must insert themselves now into the discussion of how these technologies are conceptualized and deployed, or risk being sidelined as our future environment takes shape.

INTRO-DUCTION

MARK SHEPARD

Cities are "smart" and getting smarter, we're told, as information processing capacity becomes embedded within and distributed throughout ever-broader regions of contemporary urban space. Artifacts, spaces and systems we interact with (and through) on a daily basis collect, store and process information about us, or are activated by our movements and transactions. No longer solely the visions of computer scientists, engineers or science fiction writers, these technologies increasingly mediate urban life in ways we have yet to fully appreciate, understand or even regulate. As commercial interests and law enforcement agencies begin to implement urban infrastructures imbued with the capacity to remember, correlate and anticipate, we find ourselves on the cusp of a near-future city capable of reflexively monitoring its environment and our behavior within it, becoming an active agent in the organization of everyday life. To the extent that these technologies (and how we use them) influence how we experience the city and the choices we make there, they challenge the role traditionally played by architects in shaping the urban environment, a role which has historically — with a few notable exceptions — focused predominantly on the organization of space and material in terms of built form.

This book aims to broaden the purview of architecture vis-a-vis these contemporary conditions. It presents five case studies commissioned by the Architectural League of New York for Toward the Sentient City, an exhibition I curated in the fall of 2009 that critically explored the evolving relations between ubiquitous computing, architecture and urban space. These commissioned projects were selected from 152 submissions received from 22 countries in response to an open call for proposals. Building on a discourse initiated by the Architecture and Situated Technologies symposium at the League in the fall of 2006 that I organized together with Omar Khan and Trebor Scholz, and extended through the Situated Technologies Pamphlets Series that we continue to edit, the exhibition attempted to manifest some of the more abstract concepts that have evolved out of a series of trans-disciplinary conversations between researchers, writers and other practitioners of architecture, art, philosophy of technology, comparative media study, performance studies, and engineering. Drawing on this rich discourse and the questions it helped to articulate, the material presented in this book proposes new conceptual models, sites of practice and working methods for an expanded field.

In my essay "Toward the Sentient City," I introduce Archigram's 1963 exhibition Living City as both a point of departure and contrast for the Sentient City project. In examining how the ways we conceive, organize and interact within urban environments have evolved since then, I argue for a reconsideration of the role of architects and the profession of architecture in relation to more recent technological developments. The chapter addresses the implications of higher-order information processing in urban environments, and examines various ways in which the concept of "sentience" when applied to non-human actors (such as cities) produces a rupture within an historical continuum

Toward the Sentient City, **The Architectural League of New York, 2009. Photo: David Sundberg/Esto.**

that has for some time defined both the nature of cities and who (or more precisely *what*) constitutes its citizens.

Hadas Steiner's essay "Systems, Objectified" delves deeper into the work of Archigram, Living City and related exhibitions Information at Museum of Modern Art and Software at the Jewish Museum, both mounted in 1970. Through revisiting works in these exhibitions, Steiner shows how the idea of architecture as an environmental condition — where space is understood more through interaction than delineation — remains no less antithetical to the traditional mores of modern design than it was half a century ago, and suggests that who or what has control over the environment remains the very essence of what is at stake.

The five case studies present documentation of the projects commissioned for the exhibition, combined with that of preliminary experiments, background research, material investigations, interaction prototypes, and the subsequent analysis and interpretation of their implementation.

Submerging the Sentient City under water, The Living (David Benjamin and Soo-in Yang) in collaboration with Natalie Jeremijenko of the xDesign Environmental Health Clinic at New York University present *Amphibious Architecture*, a public interface to water quality and aquatic life of urban rivers, and our interest therein. Two networks of floating interactive tubes, installed at sites in the East River and the Bronx River, house a range of sensors below water and an array of lights above water. What new interaction partners for environmental governance might ubiquitous computing enable? The project encourages us to expand our view of what constitutes the city and its citizens, aiming to spark public interest in and discussion about shared urban ecologies.

Natural Fuse, by Haque Design+Research, is a city-wide network of electronically-assisted plants that act both as energy providers as well as a shared 'carbon sink' resource. The project presents a coercive strategy for collective cooperation in regulating energy consumption through a network of natural 'circuit breakers' distributed throughout the city. If people cooperate on energy use, the plants thrive (and everyone may use more energy); but if they don't, the network starts to kill plants, thus diminishing the network's electrical capacity. At a time when the enclosure of public amenities by private interests has become commonplace, Haque deploys a more horizontally distributed notion of "the commons" prompting us to rethink our relationship to shared resources.

Projecting a future scenario where ubiquitous technologies help make 100% recycling a reality, *Trash Track*, by the SENSEable City Lab at MIT, deploys a set of smart tags on different types of trash and follows these through the city's waste management system to reveal the end-of-life journey of our everyday objects. The project examines what kind of agency objects might gain once given the ability to self-report their end of life-cycle conditions. Asking what happens when our waste is no longer "out of sight, out of mind," the project makes visible the invisible infrastructures of trash removal, promoting a bottom-up approach to managing resources and promoting behavioral change.

Too Smart City, by David Jimison and Joo Yoon Paek, posits a set of "intelligent" street furniture that behaves in unexpected ways, each embedded with computing and robotic systems to augment their respective roles in public space. The project explores technological failures as an attempt to highlight our shared assumptions and anticipations for how our devices, artifacts and systems should act. We generally assume our machines will treat us benignly, or at least with calculated indifference. Jimison and Paek give voice and character to these common urban objects, asking us to consider what design criteria we need for our future interactions with smart artifacts deployed by municipal agencies.

Breakout!, by Anthony Townsend and the Breakout! team, is a festival of work in the city that explores the dynamic possibilities of a single question: what if the entire city was your office? *Breakout!* posits alternative venues for collaborative work outside of traditional office buildings by injecting lightweight versions of essential office infrastructure into urban public space and coordinating impromptu meet-ups through social networking software. Countering the idea that mobile and networked technologies contribute to the social atomization of public space, *Breakout!* works toward enabling specific socio-spatial practices as a means to reinvigorate public space and stimulate collaborative work interactions.

Complementing these case studies, a series of critical essays reflect on the larger implications of ubiquitous computing for the way we think about architecture in general and the design of urban space in particular. These critical reflections help to situate what may at first appear to be novel questions about the intersection of ubiquitous computing, architecture and urban space in terms of more longstanding and established discourses on the technological mediation of urban life and the role architects, urban designers, and technologists alike might play in shaping its evolution.

Keller Easterling's essay "The Action is the Form" outlines an alternate habit of mind for architecture and urbanism with respect to the problem of form, one based on notions of the *dispositional* as articulated in the writings of Ryle, Latour and Bateson. This focus on the agency of digital infrastructures is continued with "Interaction Anxieties," by Omar Khan, which addresses the shift of agency from humans to ensembles of humans and machines, and questions our place in an emerging "Internet of Things." That the evolving ecology of people, things and digital infrastructures has spatial implications is clear. How architects and designers might apprehend and engage these new spatial conditions is perhaps less apparent.

Dan Hill's essay "New Spatial Intelligence, or the tree allowed to grow freely, but to man's pattern" questions the turn in the profession of architecture away from a basis in the capacity to think spatially in three and four dimensions and toward understanding space as simply what results from material construction. He recalls Bernard Tschumi's assertion that architecture is primarily a form of communication (of knowledge, of ideas) that is not circumscribed by the production of buildings, and evaluates the projects presented in this volume in terms of their contributions toward outlining a new spatial intelligence for architecture. In "Boxes Towards Bananas, dispersal, intelligence and animal structures," Matthew Fuller studies the spatial intelligence of non-human actors such as bees, birds, spiders and monkeys as a means to interrogate the tensions in both architecture and media systems, as described by Reyner Banham and Marshall McLuhan, between space making as an act of concentration and dispersion. Suggesting that spatial intelligence is species-specific, Fuller asks "how is it possible to imagine and to develop architectures, urban designs and modes of thought about cities that take part in realizing the question of intelligence as implicated and involved in differing spatial dynamics?"

Kazys Varnelis and the Network Architecture Lab at Columbia University Graduate School of Architecture, Planning and Preservation discuss the abstract space of the process of financialization, and offer a cautionary tale. Their essay "Space, Finance and New Technologies" builds on the social and spatial theories of Henri Lefebvre and Georg Simmel and looks toward global financial markets and their liquidation of architecture through real estate investment trusts, mortgage-backed securities, and credit default swaps for a clue to what the coming Sentient City could become: the 21st-century version of a ghost town, where most interactions and exchanges are automated by algorithms. In "Unsettling Topographic Representation," Saskia Sassen identifies a crisis in the topographic representation of cities when processes of globalization and digitization are taken into account. Suggesting that cities are increasingly assemblages of non-topographic and non-urban components, she examines how digital networks are contributing to the production of new kinds of interconnections underlying what appear as fragmented topographies, whether at the global or at the local level. Martijn de Waal focuses on the implications of these new interconnections for notions of public sphere in "The Urban Culture of Sentient Cities: From an Internet of Things to a Public

Sphere of Things." Trebor Scholz's essay "Your Mobility for Sale" introduces the concept of "geospatial labor" and provides rich examples of how an Internet of Things can be leveraged by large corporations to monetize the data we generate in the course of our daily lives. In "Comforts, Crisis, and the Rise of DIY Urbanism," Mimi Zeiger looks at how smaller-scale DIY (do it yourself) cultures have evolved since the American Institute of the City of New York founded its Laboratory for Young Scientists in 1928, and provides a reading of some of the case studies presented here as a form of DIY urbanism. Finally, social anthropologist and design researcher Anne Galloway illustrates how future-oriented expectations begin to do things today and reflects on how the case studies presented herein highlight current expectations that stand to shape the "sentient city" of the future.

Taken together, the case studies and essays that comprise Sentient City: Ubiquitous Computing, Architecture, and the Future of Urban Space are less concerned with projecting near-future urban conditions than providing concrete examples in the present around which to organize a public debate on just what kind of future we might want. At a time when environmental concerns are paramount in the minds of many, what new interaction partners for environmental governance might we identify? What different approaches might we take for working with (and within) urban ecosystems assembled by both human and non-human actors? If our infrastructure is to become capable of not just sensing the conditions within which it is operating but also perceiving something about those conditions, how do we make sure it *cares* about the way it responds? When urban artifacts, such as articles of trash, are imbued with the capacity to self-report their current location, condition and status, how will we view their new-found authority? How will we reconcile the inevitable conflicts in reporting between these artifacts and, say, those of a municipal agency? Finally, if the meaning of urban public spaces is as much a product of their spatial and material arrangement as it is of the conditions of their use, what new types of activity can be enabled in these spaces, and toward what ends?

Ultimately, Sentient City argues against a techno-determinism that cedes overwhelming agency to new technologies and either champions or laments their projected impact on urban life. Rather, the book examines the relationship between ubiquitous computing, architecture and the city in terms of the active role its citizens might play as designers, users, inhabitants and participants in the unfolding techno-social situations of near-future urban environments.

TOWARD THE SENTIENT CITY

MARK SHEPARD

Living City, *Gloop 4 Communications in the Living City,* Warren Chalk + Ron Herron, © Archigram 1963.

When it is raining in Oxford Street the architecture is no more important than the rain, in fact the weather has probably more to do with the pulsation of the Living City at that given moment.
— Peter Cook [1]

One could argue that this provocation by British architect Peter Cook remains relevant to the design and inhabitation of the contemporary city. Appearing in the introduction to an issue of Living Arts magazine published in 1963, it suggests that architecture, at least as it was traditionally conceived, might no longer play a vital role in shaping the urban experience. This issue of Living Arts was a catalog for Living City, an exhibition organized by the young British architecture collective Archigram and presented at the Institute for Contemporary Art (ICA) in London. At the time of the exhibition, "swinging" London epitomized the modernization of British cities in the 1950s and 1960s.[2] The glare of neon lights, the proliferation of urban advertising, the glitter and glam of new (American) products displayed in storefront windows, or the horror of "garishly decorated restaurants"[3] — this illuminated "pop" city became the curse of "proper" British architects and urban planners. The urban vernacular, Living City claimed, made fussing with the detailing of urban facades or interior lobbies irrelevant, as the experience of the street was more influenced by ambient, immaterial, and kinetic forces than by the detailed formal articulation of space and material.

Instead, as the exhibition made clear, the flotsam and jetsam of everyday urban life were to become the new materials of architecture. Organized around seven thematic sections — Man, Survival, Crowd, Movement, Communication, Place, and Situation— the exhibition catalogued the more ephemeral and intractable qualities of urban life, celebrating a new urban reality with which many within the architectural establishment were not entirely comfortable.[4] "We are in pursuit of an idea, a new vernacular," writes Archigram member Warren Chalk in the catalogue, "something to stand alongside the space capsules, computers and throw-away packages of an atomic/electronic age."[5] In embracing a pop aesthetic common to an everyday urbanism, Living City arrived on the London architectural scene as a provocation: architecture would need to embrace the ebb and flow of modern urban life, at least if it wanted to maintain any claim to cultural relevance.

However radical they may have seemed at the time, these claims were neither new nor original. Futurists such as Antonio Sant'Elia had long since cited the visual speed and aural volume of the modern industrial metropolis as inspiration for radical architectural forms. Likewise, urban sociologist Georg Simmel explored the impact of the visual cacophony of urban environments on the mental life of the people who lived in rapidly-industrializing cities like Berlin at the beginning of the 20th century, [6] and German cultural theorist Walter Benjamin addressed the reception of architecture as something appropriated in a state of distraction through habitual use over time in his 1936 essay "The Work of Art in the Age of Mechanical Reproduction."[7]

1. Peter Cook, Introduction to Living Arts 2 (London: Institute for Contemporary Arts and Tillotsons, 1963), 70.
2. Simon Sadler, Archigram: Architecture without architecture (Cambridge: MIT Press, 2005).
3. Christopher Booker, C. "The neo-philiacs: A study of the revolution" in English life in the fifties and sixties (London: Collins, 1969), 269. As quoted in Sadler.
4. For more background on the Living City exhibition, see Hadas Steiner, "Brutalism Exposed: Photography and the Zoom Wave," Journal of Architectural Education 59:3 (2006): 15-27.
5. Warren Chalk, "Gloop 7 Situation", in Living Arts 2 (London: Institute for Contemporary Arts and Tillotsons, 1963), 110.
6. Georg Simmel, "The Metropolis and Mental Life," in Simmel on Culture: Selected Writings (New York: Sage, 1997).
7. Walter Benjamin, "The Work of Art in the Age of Mechanical Reproduction," in Illuminations: Essays and Reflections (New York: Shocken, 1969), 217-252.

Declaring a crisis in architecture in the face of contemporary conditions as a means of arguing for its reinvention is a classic and well-worn tactic. Thirty years after Living City, Rem Koolhaas introduced the concepts of *Junkspace* and the *Generic City* to describe the global, undifferentiated extension of built space, where the drive toward Bigness superseded the attentive detailing of architectural and urban design. Declaring "The city is no longer. We can leave the theatre now,"[8] he echoes Archigram's assessment that the 'old' tools, techniques and obsessions of architecture are no longer relevant to current conditions. "People can inhabit anything," he claims. "And they can be miserable in anything and ecstatic in anything. More and more I think that architecture has nothing to do with it."[9]

Yet unlike Koolhaas, Archigram viewed the flotsam and jetsam of urban life not as something beyond the reach of architecture; rather, they sought to bring architecture beyond itself in order to engage the ephemeral qualities constituting the Living City. These were the materials of a new architecture, an urban dynamic composed of light, sound, and other forms of urban communications: "static communications + motile communications + verbal and non-verbal communications + signs + symbols," lists one montage created for the exhibit, "watch it happen + listen to the sound + see it flow."[10] Not satisfied with viewing the architecture of the city as a collection of formal, static, immobile, and timeless monuments within and around which life is organized, Archigram called for an urban architecture capable of engaging the less determinate, more ephemeral, untidy ebbs and flows of urban life: "Situation, the happenings within spaces in the city, the transient throw-away objects, the passing presence of cars and people," writes Chalk, "are as important, possibly more important, than the built demarcation of space."[11]

As the data clouds of the twenty-first century descend on the streets, sidewalks, and public spaces of contemporary cities, we might ask: to what extent are these informatic weather systems becoming "as important, possibly more important" than the formal organization of space and material in shaping our experience of the city? Since the 1980s, computer scientists and engineers have been researching ways of embedding computational "intelligence" into the built environment. Looking beyond the paradigm of personal computing, which placed the computer in the foreground of our attention, research in ubiquitous computing projected a world where computers would disappear into the background, displaced to the periphery of our awareness. Enabled by tiny, inexpensive microprocessors and low-power wireless sensor networks, processing was to become ambient. No longer solely "virtual," human interaction with and through computers in this near-future world would be more socially integrated and spatially contingent as everyday objects and spaces became linked through networked computing.

Today, as computing leaves the desktop and spills out onto the sidewalks, streets and public spaces of the city, we increasingly find information processing capacity embedded within and distributed throughout the material fabric of everyday urban space. On any given day, we pass through transportation systems using magnetic strip or Radio Frequency ID (RFID) tags to

8. Rem Koolhaas, "Generic city," in S,M,L,XL (New York: The Monacelli Press, 1995).
9. Katrina Heron, "From Bauhaus to Koolhaas," in Wired Magazine issue 4.07 (July 1996), http://www.wired.com/wired/archive/4.07/koolhaas.html
10. Sadler, 56-57.
11. Chalk, 110.

Bill Sullivan. ***More Turns*****, 2004. Courtesy of Bill Sullivan.**

pay a fare; we coordinate meeting times and places through SMS text messaging on the run; we cluster in cafes and parks where WiFi is free; we move in and out of spaces blanketed by CCTV surveillance cameras monitored by computer vision systems. Artifacts and systems we interact with daily collect, store and process information about us, or are activated by our movements and transactions.

Ubiquitous computing evangelists herald a coming age of urban infrastructure capable of sensing and responding to the events and activities transpiring within the city. Imbued with the capacity to remember, correlate and anticipate, this near-future "sentient" city is envisioned as being capable of reflexively monitoring its environment and our behavior within it, becoming an active agent in the organization of everyday life in urban public space. Few may quibble about "smart" traffic light control systems that more efficiently manage the movement of cars, trucks, and busses on our city streets. But some may be irritated when discount coupons for their favorite espresso drink are beamed to their mobile phone as they pass by Starbucks. And many are likely to protest when they are denied passage through a subway turnstile because it "senses" that their purchasing history, mobility patterns and current galvanic skin response (GSR) reading happen to match the profile of a terrorist.

That these evolving urban conditions alter traditional sites of practice and working methods of architecture and urban planning is obvious. Less apparent is how these fields might respond — indeed, even influence — the development trajectory of urban environments. While comparisons of contemporary conditions to those of Archigram's Living City do help situate these problems in terms of historical cycles of crisis and renewal in professional disciplines, both the nature of cities and the technologies enabling them have evolved significantly since 1963. In particular, two threads can be outlined that help reframe the contemporary city in terms of the specific techno-social spaces it presents and the organizational logics underlying them. The first concerns looking beyond materiality in architecture and shifts the locus of practice from the architectural "hardware" of urban space to the immaterial architecture of "software" infrastructures and their ability to inform, perform and enact new urban organizations and experiences. This thread extends ideas introduced by Living City and establishes an historical continuity with current conditions of urban life. The second addresses the implications of higher-order information processing in urban environments, and examines various ways in which the concept of "sentience" when applied to non-human actors (such as cities) produces a rupture within an historical continuum that has for centuries defined both the nature of cities and who (or more precisely what) constitutes its citizens.

Beyond Materiality in Architecture

The modern city exists as a haze of software instructions. Nearly every urban practice is mediated by code.
—Amin and Thrift [12]

Two years prior to Living City, author and urban activist Jane Jacobs published her influential book The Death and Life

12. Ash Amin and Nigel Thrift, Cities: Reimagining urban theory (Cambridge: Polity Press, 2002), 125.

of Great American Cities. In one extended passage, she describes the cycle of daily (and nightly) activity transpiring on Hudson Street, located in her neighborhood on the Lower West Side of Manhattan. This narrative of a sidewalk "ballet," as she calls it, takes the form of a list of casual events, encounters, and interactions between neighbors, workers, and passers-by:

Mr Halpert unlocking the laundry's handcart from its mooring to a cellar door, Joe Cornacchia's son-in-law stacking out the empty crates from the delicatessen, the barber bringing out his sidewalk folding chair... Simultaneously, numbers of women in housedresses have emerged and as they crisscross with one another they pause for quick conversations that sound with either laughter or joint indignation, never, it seems, anything between... Longshoremen who are not working that day gather at the White Horse or the Ideal or the International for beer and conversation... As darkness thickens... the ballet goes on under lights, eddying back and forth but intensifying at the bright spotlight pools of Joe's sidewalk pizza dispensary, the bars, the delicatessen, the restaurant and the drug store.
— Jane Jacobs [13]

If Archigram viewed the city as an environment that conditions our emotions, today, the "feel" of the street is defined less and less by what we can see with the naked eye. Taking a cue from Jacobs, Dan Hill describes the informational ballet transpiring on a typical street today in terms of what we cannot see:

We can't see how the street is immersed in a twitching, pulsing cloud of data... This is a new kind of data, collective and individual, aggregated and discrete, open and closed, constantly logging impossibly detailed patterns of behaviour. The behaviour of the street. Such data emerges from the feet of three friends, grimly jogging past, whose Nike+ shoes track the frequency and duration of every step, comparing against pre-set targets for each individual runner. This is cross-referenced with playlist data emerging from their three iPods. Similar performance data is being captured in the engine control systems of a stationary BMW waiting at a traffic light, beaming information back to the BMW service centre associated with the car's owner. The traffic light system itself is capturing and collating data about traffic and pedestrian flow, based on real-time patterns surrounding the light, and conveying the state of congestion in the neighbourhood to the traffic planning authority for that region, which alters the lights' behaviour accordingly... In an adjacent newsagent's, the stock control system updates as a newspaper is purchased, with data about consumption emerging from the EFTPOS system used to purchase the paper, triggering transactions in the customer's bank account records. Data emerges from the seven simultaneous phone conversations (with one call via Skype and six cellular phones) amongst the group of people waiting at the pedestrian crossing nearest the newsagent.
— Dan Hill [14]

To understand the implications of this folding of people, street, and data onto each other requires thinking about space in non-visual ways, where formal geometry and material articulation become less relevant than the topologies of networked information systems and their intersection with the socio-spatial practices of daily life. Martin Dodge and Rob Kitchen have suggested that

13. Jane Jacobs, The Death and Life of Great American Cities (New York: Random House, 1961), 52-53.
14. Dan Hill, "The Street as Platform," City of Sound, 11 February, 2008. http://www.cityofsound.com/blog/2008/02/the-street-as-p.html (last accessed 5 July 2010).

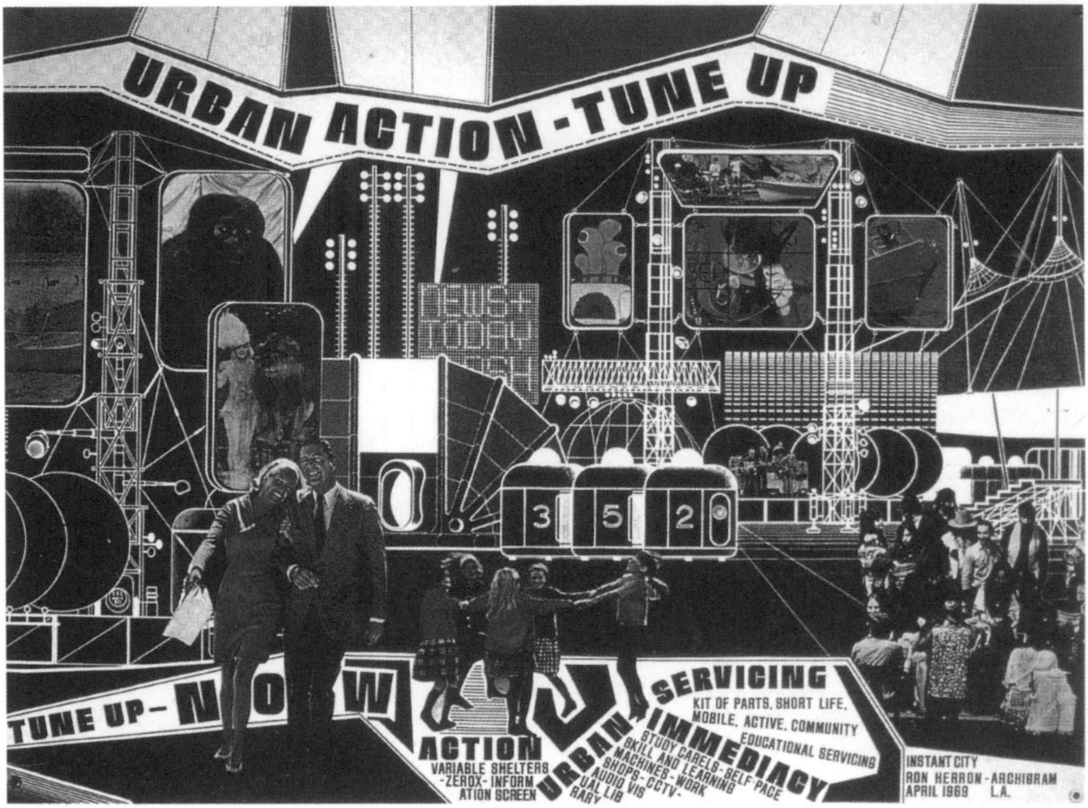

Archigram. Instant City. Urban Action Tune-Up. Collage. 1969. Courtesy Simon Herron.

these kinds of "code/space" need to be understood ontogenetically, that is, as something continually brought into being through specific practices that alter the conditions under which space itself is (re)produced. [15] Building on the work of Adrian MacKenzie, [16] they differentiate between technicity (the productive power of technology to make things happen) and its realization through transduction (the constant making anew of a domain in reiterative and transformative practices). [17] These assemblages of code, people, and space are thus brought into being through specific techno-social performances or enactments within the course of daily life.

The idea that space itself is a social product [18] — one less designed and constructed than enacted or performed through specific behaviors and practices [19] — runs counter to the absolute ontologies of space based on the Cartesian grid and Euclidean geometry that underlie the dominant spatial models of mainstream 20th century architecture. To a certain extent, Living City launched a frontal assault on these models, and Archigram explored alternatives to these spatial absolutes through a series of projects that followed the Living City exhibition. From Plug-in City, which proposed a mega-structural framework to be in-filled by its inhabitants as they saw fit, to Instant City, a kit of parts and set of procedures for delivering temporary entertainment infrastructures to sleepy English suburbs, Archigram looked for ways by which the technicity of architecture would, at least in part, be dependent on its transduction through the behavior and practices of its inhabitants.

Yet if these projects sought to open up the material 'hardware' of architecture to change and adaptation over time in the context of varied uses, others pursued 'software' architectures or spatial 'programs' as a way to redress the spatial absolutism of high modernism. [20] An architectural program is a set of functional requirements or uses ascribed to a set of spaces that, when taken together, make up a building. A program for a house, for example, typically consists of a number of bedrooms, a living room, kitchen, dining room, and supporting spaces such as bathrooms, hallways, etc. Each of these named spaces is associated with a set of normative activities or uses that they are designed to support. Encoded within the architectural program, then, are a set of behaviors and activities that are given form through the material support of designed furnishings, fittings and spatial arrangements. At the scale of the city, this takes the form of planning and zoning codes governing land use and of legislation regulating what one can and cannot do in urban public space.

One response to programmatic absolutes, perhaps illustrated best by Koolhaas, is that of the 'generic' program. The generic program recognizes that in certain instances, people will inhabit and make use of any space whatsoever — regardless of, indeed even in spite of, its formal arrangement and functional assignment. Take the conversion of industrial loft space to multi-million dollar residences, for example. Here, the form of the spatial container remains the same, regardless of its function. Another response, perhaps best represented by Bernard Tschumi, involves deliberately planning programmatic hybrids and disjunctions that provoke new uses and activities that are, by design,

15. See Dodge and Kitchin, op. cit.
16. See for example Adrian Mackenzie, Transductions: Bodies and machines at speed (London: Continuum Press, 2002) or Transduction: Invention, innovation and collective life, Unpublished paper (Lancaster: Institute for Cultural Research, Lancaster University, 2003) http://www.lancs.ac.uk/staff/mackenza/papers/transduction.pdf (last accessed 5 July 2010).
17. Martin Dodge and Rob Kitchin, "Code and the transduction of space," in Annals of the Association of American Geographers 95 (1) (2005): 162–180.
18. Henri Lefebvre, The Production of Space (London: Blackwell, 1991).
19. Michel de Certeau, The Practice of Everyday Life (Berkeley: University of California Press, 1984).
20. See Steiner, op. cit.

indeterminate. Consider the popular bar/laundromat hybrid, for example. Here, the juxtaposition of ordinary programs in an unconventional way generates spatial uses and activities that break the normative roles ascribed to each program separately. The architecture becomes a stage-set for a set of spatial practices that, in turn, enact the architecture.

The Mobile Device as Territory Machine

Imagine Hegel, Marx and McLuhan encountering the keitai [mobile phone] of the twenty-first century. Georg Hegel is astonished at seeing the spirit of the era dwelling persistently in our palms. Karl Marx complains that it is an alienating fetish object. Marshall McLuhan, his eyes sparkling, chimes in that it will turn the whole world into a village—no, a house. But in the next moment, he comes upon a realization that appalls him. "But wait!," he exclaims. "My wife and children will have the equivalent of a private room with a twenty-four-hour doorway to the outside world, fully equipped with a TV, a bed, and even a bathroom. Where would my place be in such a house?"

— Kenichi Fujimoto [21]

A new twist to the old problem of 'programming' space arises with contemporary everyday practices involving mobile devices and wireless information systems. In Japan, for example, the mobile phone has been described by Kenichi Fujimoto as a personal "territory machine," capable of transforming any space — a subway train seat, a grocery store aisle, a street corner — into one's own room and personal paradise. Born out of the so-called girl's pager revolution of the 1990s, the mobile phone became a key weapon in a young Japanese girl's arsenal for waging gender warfare against older "raspy and thick-voiced" *oyaji*, intent on peeping at young female bodies from behind a newspaper. Armed with her *keitai*, speaking freely in a high-pitched voice, "wearing loose socks and munching snacks," these *kogyaru* "couldn't care less if a subhuman *oyaji* peeked at their underwear or eavesdropped on their conversations."[22] These techno-social practices remade space in the Japanese city in new ways, transforming the paternalistic communities of city streets and subway cars into private territories for women and children.

In the West, spatial practices involving the iPod are, perhaps, more familiar. Michael Bull has studied how people use these devices to mitigate the contingencies of daily life. On one level, the iPod enables one to personalize the experience of the contemporary city with one's own music collection. When you are on the bus, at lunch in the park, or shopping in the deli, the city becomes a film for which you compose the soundtrack. The iPod also provides gradients of privacy in public places, affording the listener certain exceptions to conventions for social interaction within the public domain. Donning a pair of earbuds grants the wearer a certain amount of social license, enabling one to move through the city without necessarily getting too involved and, to some extent, absolving one from responsibility to respond to what is happening around him or her. Some people use earbuds to deflect unwanted attention, finding it easier to avoid responding because they look already occupied. Faced with two people on the sidewalk, we will likely ask the one without earbuds for directions to

21. Kenichi Fujimoto, "The third stage paradigm: Territory machines from the girl's pager revolution to mobile aesthetics," in Personal, Portable, Pedestrian: Mobile Phones in Japanese Life, ed. M. Ito, D. Okabe, and M. Matsuda (Cambridge: MIT Press, 2006), 77.
22. Fujimoto, 98.

the nearest subway entrance. In the same way, removing one's earbuds when talking to someone pays the speaker a compliment. So, in effect, the iPod becomes a tool for organizing space, time, and the boundaries around the body in public space.

What is significant here is that as these mobile devices become ubiquitous in urban environments (and in many places they already are), the technicity of architecture as the primary technology of space making is challenged by the spatial transductions these devices afford. Regardless of the formal geometries and material arrangements of a space as defined by architecture, and irrespective of the normative activities or uses encoded (or elicited) by its program, these devices and the ways in which we use them have perhaps become as important as — if not more important than — architecture in shaping our experience of urban space.

Urban Computing, Locative Media and the Read/Write City

The relatively recent emergence of Location-based Services (LBS) for mobile devices is beginning to provide insights into new ways by which information can be accessed, shared and distributed in urban environments. These services deliver information specific to location, time of day/year, and the preferences or profile of an individual, group or network. Common commercial applications include GPS-based navigation systems for mobile phones that direct one to a local restaurant or business meeting, and location-aware city guides that assist visitors to unfamiliar places in finding what they are looking for. But location-based services can also enable ordinary people to tie bits of media and information to specific locations in the physical world, marking up the built environment with personal notations, stories and images. Less common are applications that explore the ability of GPS enabled technologies to correlate place, time and identity in ways that empower users to not just read information, but also write it.

With Apple's introduction of the iPhone 3G in the summer of 2008, urban computing and locative media — formerly indicating somewhat experimental research or artist-driven explorations of location-based technologies — began being mainstreamed to the masses. We are now beginning to see social practices emerge by which location-based or context-aware media and information are consumed *en-masse* in urban environments, and, in turn, how urban space is transduced in the process.

For one thing, the way we read the city is changing. As Varnelis and Meisterlin note:

As we have grown accustomed to navigating the city with our smartphones and our printouts from Google maps, we have come to know it from above, as a two-dimensional, planimetric experience. Instead of seeing ourselves as part of the city fabric, inhabiting a three-dimensional urban condition, we dwell in a permanent out-of-body experience, displaced from our own locations, seeing ourselves as moving dots or pins on a map.

— Varnelis and Meisterlin [23]

Here, the emphasis on reading the city through "intelligent" maps, and on the implications for urban experience of the habitual

23. Kazys Varnelis and Leah Meisterlin, "The invisible city: Design in the age of intelligent maps," Adobe Design Center Think Tank, http://www.adobe.com/designcenter/thinktank/tt_varnelis.html (last accessed 5 July 2010).

patterns by which we use them, implies forms of passive consumption with which we are all likely familiar. The city becomes a network of nodes and pathways through which we circulate like data packets. "The city is here for me to use" is the underlying logic: a searchable city with an easily accessible shopping cart. [24] If, as McLuhan suggested[25], every extension of our capabilities leads to a corresponding amputation of another, wayfinding skills grounded in physical geographies run the risk of atrophy in an age of intelligent maps. Reports of mishaps stemming from reliance on GPS-enabled SatNav devices are becoming common. Recently, the London Daily Mirror reported that these devices have been responsible for at least 300,000 accidents, including that of Paula Ceely, 20, of Wales, who "vowed never to listen to her SatNav again after she was directed into the path of a speeding train at the Ffynnongain level crossing in Wales. The train slammed into her car, leaving the student within inches of her life."[26] Fortunately, no one was hurt in this instance.

If location-based technologies such as GPS navigation systems can lead to both a disembodied experience of the city as well as potential bodily harm, this has as much to do with the ways in which we use the technologies and the practices by which this space is enacted as it does with the technology itself. In this regard, revisiting early work in locative media that focuses on urban environments is instructive. Amsterdam Realtime (2002) [27], a project by the WAAG society in association with Esther Polak, traced movements through the city of people carrying GPS-enabled devices, which transmitted their location in real time to a remote server that, in turn, projected these movements as an animated "map" in an art gallery. This map represented the city not as a static network of streets, buildings, and spaces, but as a series of traces that aggregate over time to represent the city as different people traverse it. Here, the traditional authority attributed to maps and their ability to structure the way we navigate cities is subverted. Rather than a map that informs how one moves through a city, one's movements inform the map.

Further, the ability not only to read, in situ, bits of media and information associated with specific locations in the city, but also to write or otherwise add geocoded data to these urban data clouds, leads to more subtle shifts in the way we experience the city and the choices we make within it. As Malcolm McCullough notes [28], cities have throughout history been inscribed by various information layers that shape our experience of urban space, be they "grand expressions carved in stone facades, mundane signage in the streets," or smaller markings identifying significant sites or directing traffic and pedestrian flow. These urban annotations, in the past, have been governed by various public and private agencies, defined by different communities of practice: utilities providers, tax assessors, insurance underwriters, urban historians. When open to public consumption, these markings have generally served specific private interests, such as local business improvement districts (BIDs) or community associations. One of the more significant aspects of urban computing and locative media is how they open up the process of urban annotation by enabling ordinary people to contribute to the information layer overlaid on contemporary cities.

24. It should come as no surprise that the design and development of urban informatic systems is currently dominated by people coming from a background in web design. Despite the fact that these are very smart, extremely talented people, they struggle — as we all do — with the received assumptions, latent biases, and hidden agendas that one's background inevitably brings to new and relatively uncharted territory. So you find urban systems designers that can't help but view the city as a website. The dominant business model is one heavily invested in making the city easier to use for the tourist looking for that unique lunch spot or for the hipster looking for that "serendipitous" encounter with friends close-by.

25. Marshall McLuhan, Understanding Media. (New York: Mentor, 1964).

26. See Tanith Carey, "SatNav danger revealed: Navigation device blamed for causing 300,000 crashes," Mirror.co.uk, 21 July 2008, http://www.mirror.co.uk/news/top-stories/2008/07/21/satnav-danger-revealed-navigation-device-blamed-for-causing-300-000-crashes-89520-20656554/ (last accessed 5 July 2010).

27. http://realtime.waag.org (last accessed 5 July 2010).

28. Malcolm McCullough, "On urban markup: Frames of reference in location models for participatory urbanism," Leonardo Electronic Almanac 14 (3) (2006), http://leoalmanac.org/journal/vol_14/lea_v14_n03-04/mmccullough.asp (last accessed 1 September 2008).

Amsterdam Realtime, 2003. Courtesy of Amsterdam Realtime/Esther Polak, Jeroen Kee, and Waag Society.

Yellow Arrow (2004) [29], a global public art project originating in New York, employed simple SMS messaging techniques to enable people to "markup" the physical spaces of the city with virtual "tags" or "geo-annotations"— short text messages associated with a specific spot or location (see www.yellowarrow.net). Using stickers obtained from the project website, participants affix markers bearing a unique code in the public realm. When others encounter a sticker on the street, they send the code printed on it via a text message to a particular phone number. A text message is subsequently received that contains a message left by the person who placed the sticker. In place of the ubiquitous bronze plaque providing "official narratives" affixed to the side of "significant" urban structures or spaces, Yellow Arrow provides for the unofficial annotation of everyday urban places by ordinary citizens.

Urban Tapestries (2004/2006) [30], a project by the London-based social research group Proboscis, further investigated this idea of "public authoring" of urban environments. From the project website:

Urban Tapestries investigated how, by combining mobile and internet technologies with geographic information systems, people could 'author' the environment around them; a kind of Mass Observation for the 21st Century. Like the founders of Mass Observation in the 1930s, we were interested in creating opportunities for an "anthropology of ourselves" — adopting and adapting new and emerging technologies for creating and sharing everyday knowledge and experience; building up organic, collective memories that trace and embellish different kinds of relationships across places, time and communities.

The project enabled people to create relationships between places and to associate text, images, sounds and videos with them. Using a mobile phone or PDA with wireless connectivity, project participants authored "pockets" consisting of text and media objects related to specific locations in the city. A series of pockets by a single author formed "threads" that connected these locations. Significantly, the system enabled people "not only to personally map their urban spaces, but also [to] read the maps of the neighbors and strangers who share those spaces."[31] The project thus focused less on the uniqueness of individual expressions and more on the aggregation of these personal annotations and how they form a collective representation of urban life in a particular place at a particular time.

Questions concerning attention/distraction and the influence of ambient informatics on the perceptual conditions of urban space and the cognitive states of those who live in cities are longstanding. Benjamin's oft-cited observation in "The Work of Art in the Age of Mechanical Reproduction" that architecture is primarily received collectively in a state of distraction and Simmel's discussion of the origins of the blasé attitude in his seminal essay "The Metropolis and Mental Life" are both implicated in these recent transformations. More recently Clive Thompson has described a new kind of "ambient awareness" emerging out of social web media such as Twitter and Facebook status updates.[32] Individually, these short strings of text are relatively meaningless,

29. http://yellowarrow.net/
30. http://urbantapestries.net/
31. Roger Silverstone and Zoetanya Sujon, "Urban tapestries: Experimental ethnography, technological identities and place," MEDIA@LSE Electronic Working Papers 7, http://www2.lse.ac.uk/media@lse/research/mediaWorkingPapers/ewpNumber7.aspx (last accessed 5 July 2010).
32. Clive Thompson, "Brave New World of Digital Intimacy, The New York Times Magazine, September 5, 2008, http://www.nytimes.com/2008/09/07/magazine/07awareness-t.html (last accessed 5 July 2010).

Yellow Arrow is a project created by Christopher Allen, Michael Counts, Brian House, Jesse Shapins and Counts Media Inc. Photography Jesse Shapins.

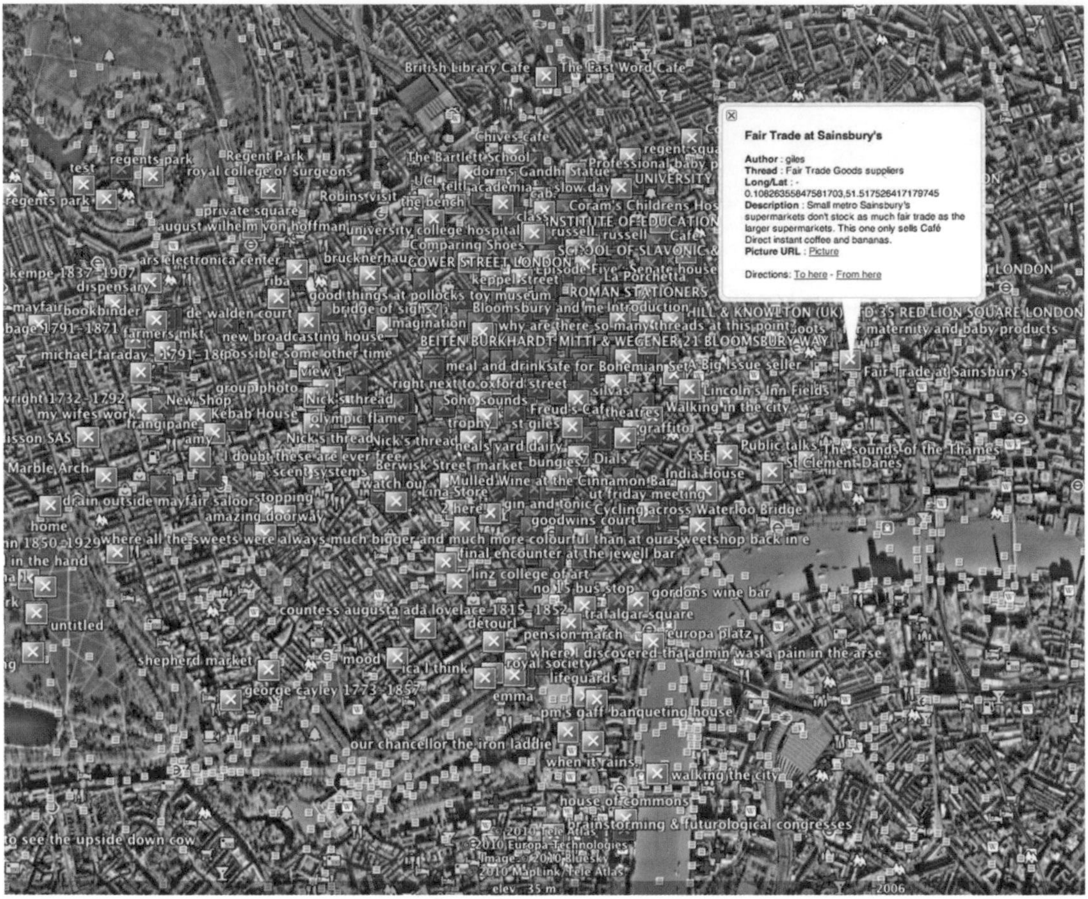

Urban Tapestries, Proboscis, 2003. Courtesy of Proboscis.

providing quotidian updates on the minutia of the daily lives of friends and acquaintances. Yet by skimming these short bits of information, Thompson suggests, we construct a peripheral awareness — a co-presence of sorts — with these absent others. As information is displaced from physical to virtual environments, our attention now becomes divided not just within our field of vision, but also between two radically different fields of vision, one human and one non-human.

But what happens when another layer of information processing is added to the mix? When we move beyond the direct action-reaction responses of systems such as subway turnstiles, or the read-write interactions involved with accessing urban markup in real-time, information processing capacity reaches a different order of magnitude. Here one's transaction history (what you've bought), mobility patterns (where you've been) and personal profile (sex, age, zipcode and related demographics) can be mined for patterns that match known profiles (of a likely customer, or a potential terrorist) and inferences can be made (what you might buy next, where you might strike). Here, computational systems operate on metadata, having been trained what to look for using neural network algorithms, where performance is measured in terms of the percentage of false-positives (or false-negatives). Here, we have urban systems and infrastructures that take on a quality of what might best be described as 'sentience' — not quite the 'smart' city we've been promised by techno-evangelists, yet not exactly dumb either.

Pathetic Fallacies and Category Mistakes: Making Sense and Nonsense of the Sentient City

And supposing there were a machine, so constructed as to think, feel, and have perception... we should, on examining its interior, find only parts which work one upon another, and never anything by which to explain a perception.
— Leibniz[33]

What are the implications of calling a city 'sentient'? The word 'sentience' refers to the ability to feel or perceive subjectively, but does not necessarily include the faculty of self-awareness. Which is to say, the possession of 'sapience' is not a necessity. Sapience can connote knowledge, consciousness, or apperception. The word 'sentience,' derived from *sentre*, means 'to feel' or 'to hear.' Sapience comes from *sapere*, meaning 'to know.' So a sentient city, then, is one that is able to *hear* and *feel* things happening within it, yet doesn't necessarily know anything in particular about them. It *feels* you, but doesn't necessarily *know* you.

Wherein lies this perception? How do we account for it? In the passage quoted from above, Leibniz goes on to claim "it is in a simple substance, and not in a compound or in a machine, that perception must be sought for." His belief that the gap between the physical and the subjective is unbridgeable, that we cannot explain subjective experience through an accounting of physical processes, can be traced to Descartes and his theory of dualism.[34] Cartesian dualism, commonly known as the "mind-body problem," asserts that mind and matter are fundamentally different

33. Gottfried Wilhelm Leibniz, Monadology, and Other Philosophical Essays (Indianapolis: Bobbs-Merrill, 1965). Originally published 1714.
34. Réne Descartes, Meditations on First Philosophy, trans. Cottingham, J., (Cambrdge: Cambridge University Press, 1966). First published 1641.

kinds of substances, and argues that mental processes are immaterial and that material organisms don't think. In Meditations on First Philosophy, Descartes attempted to account for animal behavior by purely physical processes as a means to distinguish living things that merely sense from those that are sentient. In doing so, he claims that this distinction marks an essential metaphysical difference: human beings are those that are sentient, all others are merely capable of sensing.

Sensing, the thinking goes, is something animals, some plants, and some machines can do. Sensing involves a sensing organ, or device that enables the organic or inorganic system of which it is a part, to actively respond to things happening around it. An organism or system may sense heat, light, sound, or the presence of rain, for example. Yet having a sensation or a feeling is something which goes beyond mere sensing, for it involves an internal state in which information about the environment is processed by that organism or system so that it comes to have a subjective character. *Qualia* is the philosophical term for this, which Dennett defines as "an unfamiliar term for something that could not be more familiar to each of us: the ways things seem to us."[35]

Non-human sentience has long been a flash point of controversy between the humanities and sciences. In Modern Painters, Ruskin coined the term "Pathetic Fallacy" to signify any description of inanimate things that attributes to them human capabilities, sensations, and emotions.[36] His translation of the Latin phrase *natura abhorret a vacuo* (nature abhors a vacuum) is widely known and has become part of common, everyday language. Within literature, anthropomorphism is by now an accepted literary device, yet within the natural sciences, for example, it is still considered a serious error in scientific reasoning if taken literally. Bruno Latour suggests that the difficulty lies in describing agency in the absence of anthropomorphic actors, that there is a lack of accepted vocabulary to address the non-human agency of 'things,' technological or otherwise. "[E]very time you do that," he states, "immediately people say... 'Oh, you anthropomorphize the nonhuman.' Because they have such a narrow definition of what is human, that whenever a nonhuman does something, it looks human, as if it's sort of a Disney type of animation."[37]

The term "Category Mistake" — introduced as the fundamental mistake of Cartesian dualism by Gilbert Ryle in The Concept of Mind — describes a seemingly nonsensical mixture of logics. [38] For Ryle, Cartesian dualism mistakenly assumes it is sensible to ask of a given cause, process, or event, whether it is mental or physical, implying that it cannot be both. He argues that saying "there occur mental processes" does not mean the same type of thing as saying "there occur physical processes," and, therefore, that it makes no sense to conjoin or disjoin the two. Keller Easterling elaborates on the category mistake: "For instance, one mistakes a part for a whole, or inverts levels in a hierarchy. Or a child thinks a division is a smaller part commensurate with a battalion or a squadron, when it is the overarching category for those of smaller divisions." She goes on to show how beginning with Jesus and extending to

35. Daniel Dennet, "Quining Qualia" in A. Marcel and E. Bisiach, eds, Consciousness in Modern Science (Oxford: Oxford University Press, 1988).
36. John Ruskin, Modern Painters (New York: John Wiley and Sons, 1864)
37. Bruno Latour, "Where Constant Experiments Have Been Provided," Interview, http://www.artsci.wustl.edu/~archword/interviews/latour/interview.htm
38. Gilbert Ryle, The Concept of Mind (Chicago: University of Chicago Press, 1949).

messianic characters in general, category mistakes are markers for dominant logics with universal claims. She also suggests how they can serve as an escape hatch out of the monotheisms of logic and discipline. "In order to find the trapdoor into another habit of mind, one would not quarrel with, but gather evidence in excess of" these dominant logics.[39]

The Sentient City thus becomes a contested site: a theoretical construct within which longstanding claims of essential human qualities, capabilities and characteristics are critically destabilized through their attribution to non-human actors. This destabilization is understood to work actively, as a tactical maneuver enabling other ways of thinking that not so much confront dominant ideologies but elide common wisdoms about, not only what it means to be human, but also what it might mean to be a city.

This method is, of course, by no means new. What follows is a cross-section of representations of the Sentient City culled from the fantasies of science fiction writers, the research agendas of computer scientists, and the claims accompanying recent applications deployed by corporate interests, governmental agencies, and the military. The intent here is less to provide a comprehensive overview, but rather to provender a selection of examples that point to the historical persistence and cultural pervasiveness of the sentient non-human meme.

Non-human sentience is no stranger to the science fiction community. From Arthur C. Clarke's Diaspar, the computer controlled city described in The City and the Stars, to his work with Stanley Kubrick on HAL (sentient machine); from Stanislaw Lem — and Andrei Tarkovsky's — Solaris (sentient planet) to DC Comics' Ranx, the Sentient City created by Alan Moore; from Gibson's sentient cyberspace as portrayed in Neuromancer, to the sentient programs of the Matrix, or Bruce Sterling's spime (to name but a few), science fiction has imbued a range of inanimate "things" of all scales with forms of sentience that do not map neatly to those of ordinary humans.

These technological fantasies of non-human sentience exhibit no consensus regarding the place or nature of sentience, however. Sentience is at times centralized (Clarke, Kubrick, Moore), at times distributed (Lem, Gibson, Sterling). While Clarke and Kubrick attempt to anthropomorphize HAL, as symbolized by his iconic and omnipresent red eye and reinforced by his conversational acuity, Lem persistently portrays Solaris' *otherness*: the planet's sentience is evidenced through the manipulation of a simple substance constituting its oceans that has nothing in common with anthropomorphic figuration or behavior.

Addressing sentience as a technical challenge, the Economist published an article five years ago titled "The sentient office is coming" that described then current research in augmenting computers and communication devices with sensors to enable them to take into account their environment and adapt to the changing conditions of their use.[40] Here the aim was to create "convivial technologies that are easy to live with." Yet as the article points out, cohabitation with sentient things is not without dilemmas. What happens when the toaster in your home gets bored of always making toast, or the fax machine in the office thinks the tone of your fax doesn't jive with that of the firm?

39. Keller Easterling, "Only the Many," Log 11 (2008)
40. "The sentient office is coming," The Economist, June 21, 2003.

Achieving "sentience" in the domain of Artificial Intelligence (AI) is a serious research agenda with a long history. ATT/Cambridge University's Sentient Computing project (1999) [41] attempted to "combine sensors and computers to monitor resources, maintain a computational model of the world, and act appropriately." Combining sensors and computers was at the time nothing new, but the broad attempt to "maintain a computational model of the world" proved daunting. As of 2006, the project was re-focused on tracking and location systems for "sentient" vehicles and sports.

Today the emphasis is less on trying to maintain a proprietary computational model of the world, and more on using the world itself as 'model' and letting ordinary people contribute to its making. More than a few early Urban Computing and Locative Media projects focused on crowdsourcing metadata about a place by enabling people to markup and annotate digital maps with notes, images and media objects geocoded to specific locations (Urban Tapestries, Yellow Arrow, Semapedia, to name but a few). Google Maps and Google Earth have further catalyzed the collective production of these geospatial datasets. With the introduction of the GPS enabled iPhone 3G in 2008, location-based services building on these datasets began being mainlined to the masses.

Context-awareness plays a significant role in current research in sentient systems. In addition to knowing where someone is, factors such as whom they are with and what time of day it is reduces the possibility space within which inferences and predictions are made. This real-time information is correlated with historical data of someone's mobility patterns, purchasing history, personal interests and preferences (as reflected by user-generated profiles) in order to make more accurate predictions about what his or her wants and needs may currently be, or what actions s/he is likely to take next. MIT's Serendipity project, for example, draws on the real-time sensing of proximate others using Bluetooth technologies built into mobile phones to search for matching patterns in profiles of people's interests. Developed by the Human Dynamics Group at the Media Lab, the project's goal is to facilitate corporate productivity by providing a matchmaking service for workers with shared interests or complimentary needs and skills who otherwise might not encounter each other within spaces organized around the office cubicle. A typical design scenario involves one worker needing the skills of another and the system facilitating their meeting:

When we were passing each other in the hallway, my phone would sense the presence of his phone. It would then connect to our server, which would recognize that Tom has extensive expertise in a specific area that I was currently struggling with. If both of our phones had been set to "available" mode, two picture messages would have been sent to alert us of our common interests, and we might have stopped to talk instead of walking by each other.
— Nathan Eagle [42]

This project presents at least two assumptions that are worth exploring further. The first is that 'matchmaking' should be based on comparing profiles and looking for 'synergies' between

41. http://www.cl.cam.ac.uk/research/dtg/attarchive/spirit/ (last accessed 5 July 2010).
42. Nathan Eagle, "Can Serendipity Be Planned?", MIT Sloan Management Review, Vol. 46, No. 1 (2004), 12.

two people. If the term 'serendipity' is understood to mean the process of finding something by looking for something else, the Serendipity project does precisely the opposite: it simply outsources the problem of finding something we are already looking for (that "expertise in a specific area that I was currently struggling with" that I have somehow indicated in my profile). Secondly, while the introduction of an "available" mode suggests that some attempt has been made to address privacy issues, there is no consideration of who has access to your profile data and how they use it.

Profile data considered private in one context can be publicly revealing in another. Another MIT project, code-named *Gaydar*, mined Facebook profile information to see if people were revealing more than they realized by using the social networking site. By looking at a person's online friends, they found that they could predict whether the person was gay. They did this with a software program that looked at the gender and sexuality of a person's friends and, using statistical analysis, made a prediction. While the project lacked scientific rigor — they verified their results using their personal knowledge of 10 people in the network who were gay, but did not declare it on their Facebook page — it does point to the possibility that information disclosed in one context may be used to interpret information in another.

Looking upstream, Mike Crang and Stephen Graham's recent essay "Sentient Cities: Ambient intelligence and the politics of urban space" does a great job of outlining how corporate and military agendas are currently driving the technological ecosystems we're likely to cohabit within the near-future. [43] Mapping the Sentient City as operative reality, they point to location-based search results and target-marketing databases storing finely grained purchasing histories as steps toward "data-driven mass customization based on continuous, real-time monitoring of consumers." Further, citing a study by the US Defense Science Board calling for a 'New Manhattan Project' based on Ambient Intelligence for "Tracking, Targeting and Locating" (DSB, 2004), they outline an Orwellian future that is in fact currently in operation in lower Manhattan.

The Lower Manhattan Security Initiative, as the plan is called, resembles London's so-called Ring of Steel, an extensive web of cameras and roadblocks designed to detect, track and deter terrorists. The system went live in November of 2008 with 156 surveillance cameras and 30 mobile license plate readers. Designed for 3,000 public and private security cameras below Canal Street, this system will include not only license plate readers but also movable roadblocks. Pivoting gates are being installed at critical intersections, which would swing out to block traffic or a suspect car at the push of a button.

While the implications of projects like Serendipity occupy a relatively benign problem space, The Lower Manhattan Security Initiative points toward possibly more serious outcomes from the false positives (or false negatives) inevitably generated by the pattern matching and data mining algorithms at the core of the system. What happens when Facebook profile data is added to the mix? How do we ensure the privacy of data about us that is collected through inference engines? What are the mechanisms

43. Mike Crang and Stephen Graham, "Sentient Cities: Ambient intelligence and the politics of urban space," Information, Communication & Society, 10:6 (2007), 789 – 817.

Recommendation:
Establish "Manhattan Project"-Like Program for TTL

- **Vision**
  - Locate, identify, and track people, things, and activities—in an environment of one in a million—to give the United States the same advantage in asymmetric warfare it has today in conventional warfare
- **Structure requires that CIA, Defense, Justice, and Homeland Security**
  - Agree this is an urgent national security requirement
  - Agree on centralized management to conduct research, acquire systems, implement architecture, manage operations, and integrate results
  - Agree on funding, legal, ethical, and jurisdictional issues
  - Agree on executive responsibility
  - Acknowledge this function as a Presidential priority

The global war on terrorism cannot be won without a "Manhattan Project"-like TTL program. Cost is not the issue; failure in the global war on terrorism is the real question.

Tracking, Targeting and Locating. From a report by the Defense Science Board, Office of the Undersecretary of Defense, Washington DC, 2004.

by which these systems will gain our trust? In what ways does our autonomy become compromised?

To the extent that business interests and government agencies drive these technological developments, we can expect to see new forms of consumption, surveillance and control emerge. Despite the obvious implications for the built environment, architects have largely been absent from this current discussion. Forty years ago Reyner Banham illustrated an architecture of the "well-tempered environment,"[44] where the conditioning of space and its attendant technologies were literally drawn out of the woodwork. Yet while advances in the design of building management systems (BMS) since then have enabled greater environmental responsiveness at the scale of a building, relatively little attention has been paid to the space between and beyond buildings — the sidewalks, streets, infrastructures and urban public spaces that give form to urban life. If one accepts that the various ways we interact with (and through) these embedded, mobile and pervasive technologies can shape our experience of the city and the choices we make there, then the role of architects in shaping these technologies becomes apparent. Architects are trained to shape our constructed environment, and are skilled in orchestrating complex relationships between space, material, technologies, and various modes of habitation and use. The critical question remains: can the profession of architecture at large engage a form of practice that no longer places the act of making buildings as the central and defining role of the architect?

The profession has a decision to make. Either it can cede the role of being the primary agent in shaping our spatial experience of the city to the designers and engineers of these technologies, or it can shed its disciplinary anxieties regarding the purview of its practice and take part in shaping these technologies. This is easier said than done, to be sure. Part of what inhibits the profession from doing so lies in the desire or need to cling to certain disciplinary claims. As with most cases of identity politics, these claims are staked in terms of power, authority, status and influence. There is a lot at stake in grounding architecture in material form, for instance, not only for architectural education (established pedagogy) but also for the construction industry and the development and finance worlds (ie: the underlying conditions of production). The ability to marshal material resources in vast quantities and at large scales has long been a measure of status and power. There were clear reasons why, for example, Vespasian built the Colosseum in Rome. For architects to address new and emerging kinds of space, they (and their clients) will need to learn to see them as valuable — as spaces open to the architectural imagination — even when they are not as heavily invested in material form appropriate to building as we know it.

44. Reyner Banham, The Architecture of the Well-tempered Environment (Chicago: University of Chicago Press, 1984).

SYSTEMS, OBJECTIFIED

HADAS STEINER

During the 1960s, the Archigram trademark sheltered an array of projects that combined theoretically (despite protests to the contrary) to cultivate a proclivity for the comprehension of space through interaction rather than delineation. The Archigram position was that a gamut of external and internal pressures, from weather to desire, conditioned the behavior of humans in space; at stake was who or what would have influence over the environment. Original readership of the Archigram publication, with its portraits of reflexive structures, as well as the audiences for its exhibitions and presentations, were predominantly composed of young architects for whom the commercial application of technological developments after WWII provided inspiration. An audience of remarkably similar composition has reinvigorated interest in the Archigram project since the rise of digital technologies to cultural dominance. And yet contemporary work that furthers the spatial argument visualized by the range of Archigram schemes still hovers some fifty years later at the fringes of the architectural discipline, moored instead to the expanding field of the media arts. The idea of architecture as an environmental condition, the Sentient City proposals demonstrate, remains no less antithetical to the traditional mores of modern design than it was half a century ago.

The first collaboration of the core members who became known as the Archigram group — Warren Chalk, Peter Cook, Dennis Crompton, David Greene, Ron Herron and Michael Webb — along with some others was an exhibition that declared the city to be a sum of its atmospheres. Living City, as the exhibition was called, took place at the Institute of Contemporary Arts (ICA), London in 1963. The group assembled a structure of triangulated panels powered by its own electronic circuit within the gallery hall. Upon entering, the visitor had the sensation of entering into a three-dimensional collage calculated to overwhelm visually and aurally, reinforced by basic technologies, such as a soundtrack, flickering light machine and periscope that enabled one to peer disjunctively outside of the exhibition at the legs of passersby. Surrounded by images, sounds and lights, the visitors were as disconnected from the gallery space as from the predictable and accepted relationships of the everyday. All this was quite unlike the standard display of drawings and models that was the norm for an architectural exhibition.

"What we really need," Warren Chalk explained, "is increased environmental stimulus. Because the environmental stimulus is weak, man is inventing novelties like wife-swapping and unisex dressing. He is bored."[1] The first Archigram antidote to such tedium was a proposed shift in focus from conventional architectural typologies to criteria based on program, such as those suggested by the governing themes of the exhibition: Man, Survival, Crowd, Movement, Communication, Place and Situation. Though all of these categories had already appeared in mainstream and avant-garde discourse, in concert they became an agenda. Each theme, as in life, was meant to interact with and adapt to the others, creating unexpected connections. The vitality of the city was to be generated, not by the demarcations of the built environment, but rather by the interactions that took place wherever people gathered. [2]

1. Warren Chalk, "Trying to Find Out is One of my Constant Doings," A Guide to Archigram 1961-74 (London: Academy Editions, 1994), p. 322.
2. Peter Cook, "Come-Go: The key to the vitality of the city," Living Arts 2, 1963, p. 80.

As such, the overall structure of the exhibition was a spatial demonstration of how the themes came together and diverged within what was described as "simulations of city life, not a display of suggested forms."[3] (see image page 16) In plan, the structure was divided into seven intersecting alcoves, or "gloops" as they were called in a manner suggestive of overlapping cybernetic loops, one for each component. Man, Crowd and Survival, themes that related to the inhabitants, formed a cluster to the left of the entrance point. These categories were contingent upon the variables of Movement, Place and Situation that unfolded in a crescent to the right. Communication, the stimulant that enabled all the social conditions for activity, had pride of place on entrance-right. The small pockets of the exhibition were akin to the bubbles of subjectivity in which individual encounters occur.

In various guises, the mutually dependent categories of Living City would sustain a decade of investigation in projects developed by the group, from the urban schemes for which they became best known, to those that explored social interactions, systems of circulation and aggregations of components. These projects, in turn, provided a visible anchor for the larger cultural effort to reconcile the entrenched view of objects, buildings included, as discrete, engineered entities with the network-based, nomadic programs. In The System of Objects (1968), the published iteration of the doctoral thesis that Jean Baudrillard wrote under the palpable guidance of Henri Lefebvre, for example, the home-dweller was defined no longer as an owner or user but as an "active engineer of atmosphere." Baudrillard explained:

Space is at his disposal like a kind of distributed system, and by controlling this space he holds sway over all possible reciprocal relations between the objects therein, and hence over all the roles they are capable of assuming. [4]

The multiplicity of practices taking on this conceptualization of environmental conditions and situations at the expense of the fixed object signified a new direction for artistic production. In July of 1970, the Museum of Modern Art opened Information, its first exhibition to recognize and institutionally acknowledge the phenomenon that had become known as conceptual art. [5] The Jewish Museum, then an adventurous venue, would follow on MoMA's coattails with Software in September of that year. [6]

Jack Burnham, the curator hired for the latter, was known for his criticism in Arts magazine and Artforum that defined the agenda of communications technology in relation to contemporary art. These essays were replete with discussions of organization, participation and entropy, in short, of time, with software theorized as a metaphor for the expanded framework of the art process. [7] The consumption of art by an audience was argued to be as fundamental a part of the system as the conception and execution of a project. With his first curatorial effort, Burnham hoped to demonstrate "software" as a parallel course to the aesthetic principles, concepts, or programs that resulted in art objects, or "hardware." Images, Burnham stated in the catalogue, were "secondary or instructional." The

3. Cook, "Introduction," Living Arts 2, p. 68.
4. Baudrillard, System of Objects (Brooklyn, NY: Verso), 2006, p. 26.
5. The exhibition had competition from another show, Conceptual Art and Conceptual Aspects, being held at the New York Cultural Center from April 10 to August 25, 1970.
6. For much more on the Software exhibition at the Jewish Museum, see E.A. Shanken, "The House That Jack Built: Jack Burnham's Concept of 'Software' as a Metaphor for Art," Leonardo Electronic Almanac 6:10 (November, 1998), http://mitpress.mit.edu/e-journals/LEA/ARTICLES/jack.html, and "Art in the Information Age: Technology and Conceptual Art," Leonardo, Vol. 35, No. 4, pp. 433 - 438.
7. "A polarity is presently developing between the finite, unique work of high art, i.e. painting or sculpture, and conceptions which can loosely be termed 'unobjects,' these being either environments or artifacts which resist prevailing critical analysis." Jack Burnham, "Systems Esthetics," Great Western Salt Works (New York: George Braziller, 1974), p. 15.

Architecture Machine Group, SEEK, 1970. The Jewish Museum, New York/Art. Resource/NY

exhibition emphasized inter-disciplinary engagement and mixed knowledge of computer science with contributors running the gamut from Robert Smithson, Hans Haacke and Nam June Paik to the Architecture Machine Group. Architecture Machine Group contributed SEEK, a gerbil populated habitat in which the program adapted to unpredictable disruptions of the environment and activated a magnetized robotic arm that configured blocks in response to the activity of the rodents. Other installations also engaged with computer technology. At the other end of the spectrum, John Baldessari cremated his "accumulated art"—all of his hardware. This exhibition, despite its conceptual sophistication, was dogged by problems that diminished its impact. Crucially, the software designed to animate the exhibits could not be made to function properly during the first half of the show, subjecting the premise itself to critique. [8] SEEK is best remembered for the havoc caused by the gerbils who turned against each other and ran amok. Jean Toche, a Belgian artist who had planned to exhibit a tunnel of polluted air, withdrew when informed that American Motors Corporation was the sponsor for the show. Moreover, the exhibition ran so severely over-budget that Burnham nearly bankrupted the museum and the director was forced to step down.

8. The programs were written to run on a PDP-8 minicomputer (the interim phase before personal ones).

By contrast, while exhibiting a cluster of the same artists as would participate in Software — Vito Acconci, Joseph Kosuth, Dennis Oppenheim, Ed Ruscha, and Robert Smithson — among other prominent (and less familiar) figures, Information veered away from risky technological deployment and towards the institutionally familiar domain of the image, no matter how grainy and intentionally unpolished, as conveyor of data. The emphasis was on art as part of a larger system of visual references, including journalistic records of tumultuous current events. What was unique about Information in the context of other exhibitions of conceptual art was that Kynaston McShine used his curatorial role to facilitate engagement, rather than to direct. The artists were allowed to represent themselves as they wished in the catalogue with work that did or did not appear in the exhibition. Hans Hollein, the sole architectural figure among the invited participants, sent site photos of non-buildings, subterranean buildings, and slight surface modifications for both the show and the publication. The curator compounded the relevance of the exhibition's message with a secondary array of captionless imagery presented in the last third of the catalogue that related to the project, whether in the form of interventions by artists or by journalists. [9] The catalogue was thus its own mode of information beyond its role as museum document.

Amongst the supplemental imagery of non-participants and alongside images of overt political content was a spread by Ant Farm adapted from Architectural Design and eight collages for "Instant City" by Ron Herron of Archigram. McShine's awareness of developments in the field of architecture relevant to his theme had likely been gleaned from his perusal of the recent Design Quarterly double-issue dedicated to "Conceptual Architecture" that was archived with the exhibition materials at MoMA, as well as copies of the Archigram magazine. [10] The guest editor for that issue, John Margolies, had also granted invitees control over their own pages as McShine would do. Peter Eisenman, invited to contribute an introduction to the subject, proffered a series of blank pages punctuated only by footnote references and the notes themselves. Ant Farm, Archigram, Archizoom, Francois Dallegret, Haus-Rucker Co, Craig Hodgetts, and Superstudio were all requested to contribute images never before published in North America. Submissions by Les Levine and Ed Ruscha served to substantiate the connection of these figures unfamiliar to the art scene to the broader conceptual agenda. In addition to duplication of Instant City collages, a letter sent by David Greene to Ant Farm further represented the Archigram cause. It read: "Dear Sirs, I read with consummate interest your spell binding literature about changing lifestyles. I have at the moment an old lifestyle which I would like to trade in for a new model." [11]

Thus in its evocation by others Instant City was doubly associated with the image/information model rather than the environment/software model that was closer to Archigram's own editorial rhetoric. But the very images that made this elision possible illuminated the tenuous relation that had always persisted between the system and its object in Archigram

9. Kynaston L. McShine, ed., Information, (NY: MoMA, 1970). For a short history of the exhibition, see Ken Allan, "Understanding Information," Conceptual Art: Theory, Myth, and Practice, Michael Corris, ed., (Cambridge, UK: Cambridge University Press, 2004), pp. 144-168.
10. Ken Allan, "Understanding Information," from Conceptual Art: Theory, Myth, and Practice, p. 148.
11. Conceptual Art, p. 9.

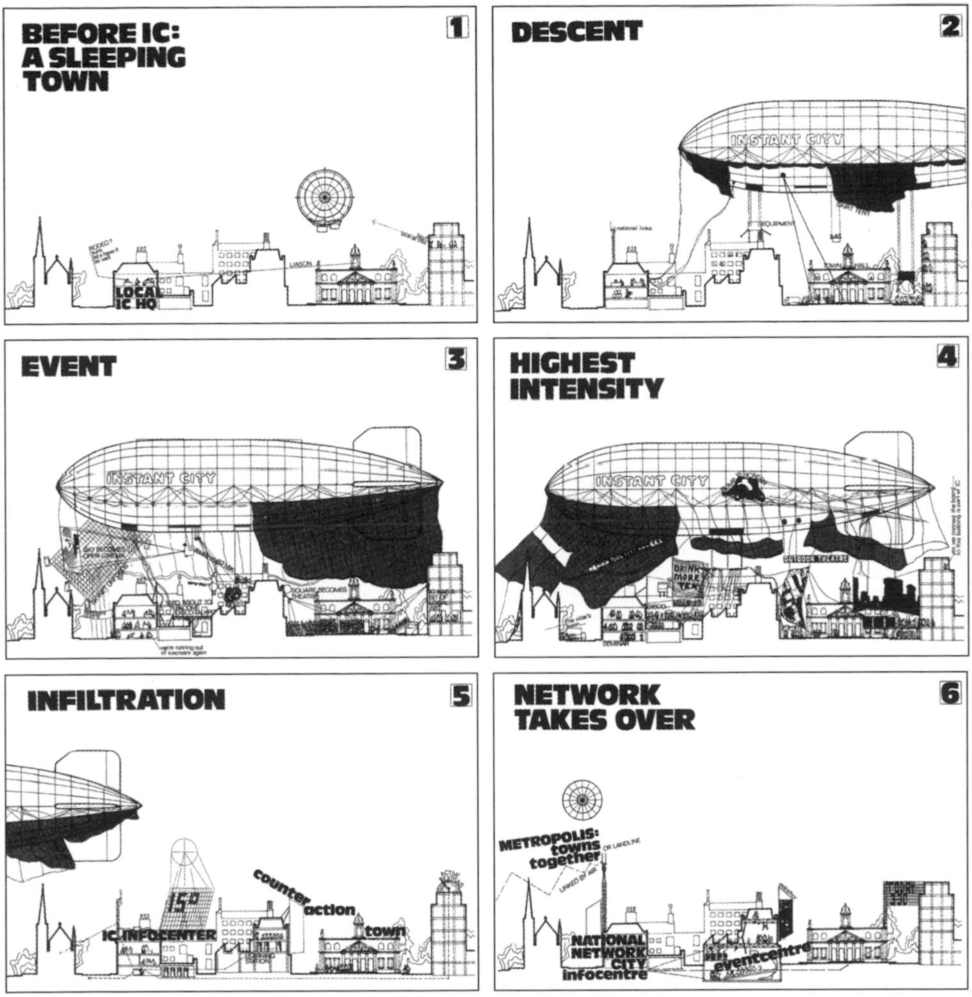

Instant City Airship, sequence of effect on a typical English Town. Peter Cook, © Archigram 1970.

projects. Unlike the most familiar Archigram proposals that were assigned single authorship, on the one hand, the group gesture of Instant City was not meant to be fixed to a single motif or image but rather composed of an array of "several on-going conversations." [12] Like the Living City that signified the beginning of group production, the instant version that was a product of its end was a joint endeavor. (see image page 22) Yet the collages for this project, with models and advertising slogans foregrounded against an architectural backdrop, would firmly associate Archigram with sixties image and consumer culture in a way that the professional axonometric, elevational and perspectival details of Plug-In, Computer or Walking Cities could not. The Archigram language had changed from categories such as Man and Situation to those of Network and Software as the sixties progressed and public awareness of computing terminology grew. Instant City sought to define the life of the city as the outcome of information flows and deposits across a greater urban network rather than the circulation of goods and services emphasized by its predecessor. And yet software often retained

12. Cook, Experimental Architecture, p. 122.

the guise of things made out of soft materials rather than fulfill the promise suggested by Dennis Crompton in Computer City.

Despite the informational assimilation of its imagery, Instant City was in fact an event: an extrapolation of the traveling circus, which synthesized the interest in transient structures, from trailer units (hardware) to pneumatic ones (software), including airships, balloons, domes, and air-supported structures, with the notion of an environmental system. The dirigible in fact served as the icon of this project. All the migrating units combined to deliver information that would alleviate the inequality and boredom of the extant provincial condition by bringing the vibrant urbanism of a major metropolis to decentralized locations. Lower density areas would be infused with services without the associated burdens of permanence and building maintenance. Components would arrive at a town by road and air and use the host city as a skeleton. Multi-media displays and information banks that substituted for educational institutions generated much of the ineffable atmosphere. The results, it was declared, tread the "theoretical territory between the 'hardware' (or the design of buildings and places) and 'software' (or the effect of information and programmation of the environment)." [13] After the experience, the hardware of Instant City moved on as quickly as it had arrived; the coming together of things already assumed their coming apart. [14] It is at this point, after the intensity of the Instant City happening has passed, that the infiltration of wiring and sensors quietly occurs. "The Network Takes Over," and the residue of information is left behind long after the image has faded. [15]

As the very symbol of the project suggested, Instant City was stuck between program and thing: "The giant, pretty, emotionally evocative object, the blimp, the airship, the beauty-and-disaster-and-history. Back to the heroic and beautiful object?" [16] We would do well to recall this self-conscious question given the renewal of interest in Archigram concerns as then speculative technologies can be implemented in practice. Instant City was, in the words of its designers, a means to an end, "a tool for the interim phase: until we have a really working all-way information network." [17] This interim stage reflected the current—then and now—limitations of the systems model, not the Archigram idea of a final destination. [18] The transience of knowledge fundamental to Instant City was evolving towards a point where information and the city were synonymous. In its ideal form, Instant City would provide a bundle of services: its urban strategy would be connectivity and speed over geographical advantage. To inhabit an advanced network, information and the city would be fully decentralized commodities that traveled the same infrastructure, like computers on phone lines. This was not a mere reduction of electrons to cables, of content to form, but an ontological reduction of the difference between structure and information. As modes of communication, both used the same methods, even more, served the same function.

For architects, the disciplinary tension between the image and object has been around since whenever it was that theory split from practice and it perseveres with evermore complexity.

13. "A Programme Background," A Guide to Archigram 1961-74, p. 246
14. "This has to be quick. It is on the scale of half an hour and five thousand people and amazement and the whetting of appetites." Architectural Design, November 1970, p. 568.
15. The final panel of six in "Instant City Airships: sequence of effect on a typical English town" (1970).
16. Architectural Design, November 1970, p. 579.
17. "Ideas Circus," A Guide to Archigram 1961-74, p. 233.
18. "Hard-Soft," A Guide to Archigram 1961-74, p. 222 (not identical to the hard-soft editorial in Archigram 8)

The editorial for Archigram 7 spoke adamantly about the state of architectural publications that had established themselves as the primary conveyor of architectural information: "THE PRINTED PAGE IS NO LONGER ENOUGH... Magazines will dissolve into hybrid networks of all media at once." [19] "Hidden Architecture," the project submitted by Superstudio to the Conceptual Architecture issue of Design Quarterly, was designed to demonstrate the vacancy of the image and what they called the "semantic redundancy" of architectural magazines. [20] In the presence of a legal witness, the members of Superstudio sealed three copies of a drawing for which the original had been destroyed in a plastic envelope that was then wrapped in a foil cover and sealed within a zinc-wrapped box in perpetuity. With the hidden scheme they produced an "architecture which is only an image of itself and our instrumental muteness." [21] If the antidote to that muteness is the dissolution of architecture and its many images into a hybrid media network, what happens in the next phase to the spaces as defined by the interactions between people, between people and things, and, not least, between things and things remains open for discussion. As media networks are embedded in and distributed throughout the city, who or what has control over the environment is the very essence of what is still at stake.

19. "International Ideas," Archigram 7, unpaginated.
20. "Conceptual Architecture," p. 54.
21. "Conceptual Architecture," p. 54.

CASE STUDIES

NEW INTERACTION PARTNERS FOR ENVIRONMENTAL GOVERNANCE

AMPHIBIOUS ARCHITECTURE

DAVID BENJAMIN & SOO-IN YANG (THE LIVING) AND NATALIE JEREMIJENKO (XDESIGN ENVIRONMENTAL HEALTH CLINIC)

Amphibious Envelopes
David Benjamin and Soo-in Yang

0. City+
The city, when boiled down to its raw ingredients, might be nothing more than flows and envelopes. The flows might involve the movement of people, plants, animals, air, water, and information — each circulating through the city in its own way. They might collide and negotiate with one another, but they might also be channeled by the city's envelopes. And the envelopes might be understood simply as various thresholds: building skins and neighborhood lines, streets and microclimate boundaries.

At this level of abstraction, it might be difficult to find solid ground or to draw conclusions about the city. But at the same time, this perspective might offer freedom to propose speculative projects and frame fundamental questions.

Who should control the city's envelopes? What are our individual and collective contributions to the city's flows? If we suppose that *in general* (city) = (flows) + (envelopes), how might we then design and build our *specific* cities in our specific times, especially in the context of ubiquitous computing and situated technologies?

1. Carbon+
As a young scientist at the end of the eighteenth century, Humphry Davy was mesmerized by the beauty and the science of burning candles. He stared in wonder for hours, of course, but also, like many of his peers, he conducted experiments. He investigated the chemical processes and isolated the organic compounds. He found that carbonic gas (carbon dioxide) played a key role in combustion and also in other natural processes. Building on the work of earlier scientists, and drawing on the outlook of his Romantic Generation, Davy linked the transformations of burning candles to the transformations of plant growth and of human and animal respiration. Each involved an exchange of carbon dioxide. Together they created a network of flows.

Davy's colleague Michael Faraday explained it this way: "*Wonderful* is it to find that the change produced by respiration, which seems so injurious to us — for we cannot breathe air twice over! — is the very life and support of plants and vegetables that grow upon the surface of the earth."[1] When humans breathed, they were not depleting a natural resource. Instead they were participating in a sustainable flow. The exchange of carbon and oxygen between animals and plants created a wonderful equilibrium.

Davy went further, expanding the loop from air to water. Through lab experiments, he demonstrated that aquatic plants produce oxygen underwater when exposed to sunlight. This dissolved oxygen is consumed by fish and other sea creatures, who, in turn, generate carbon. Aquatic plants and fish replicate in water the system of exchange that was first observed in air.

We now know that carbon and oxygen also flow *between* air and water, and we now refer to this broad system of interconnected loops as the Carbon Cycle — a network of flows involving humans, animals, plants, air, and water, in an endless and essential transformation of the planet and the city.

2. Water+
On May 15, 2000, in a rattling story about the future of global business and our planet's natural resources, <u>Fortune Magazine</u> declared, "Water promises to be to the 21st Century what oil was to the 20th Century: the precious commodity that determines the wealth of nations."[2]

Ten years later, this still seems right. Yet as a whole, our current understanding of water is patchy and our engagement of water is unimaginative. We may boast of expertise in the molecular transformations that occur in small volumes of water, as well as the civil engineering involved in managing large volumes of water, but our knowledge about ecosystems and flows of water remains murky.

When clean energy start-up Verdant Power proposed to install turbines on the bed of the East River in 2003, it faced a series of challenges due to some of these unknowns. In order to receive its permit to operate, the company was required by the New York State Department of Environmental Conservation (DEC) to investigate the fish in the area. Nobody knew how many there were, or whether the proposed turbines might kill them or scare them away. In order to find out, Verdant had to spend four years and over two million dollars. Its study eventually determined there was a healthy population of 54 species of fish, and they were not bothered by turbines. Fish could easily avoid the slow-moving blades of Verdant's propellers.

The DEC granted a permit and Verdant installed five test turbines, but the company confronted further unknowns. Verdant found that its steel and fiberglass propeller blades broke off within days of deployment. They were replaced with stronger aluminum magnesium blades, but then the bolts connecting the blades to the rotor became intolerably stressed. These problems, it turned out, were due to extremely high current velocities at the river bed. Although several prior studies had mapped the speed of the East River at its surface, nobody had known that the speed was much greater at the river bed and that it would tear apart the installed equipment.

Verdant overcame these setbacks and managed to power a supermarket and parking structure with 175 kilowatts for five months continuously. But it became clear that we still have a lot to learn about the forces of the river, the movement of fish, and other exchanges in the complex water ecosystems of our cities. In terms of flows and envelopes—and in terms of situated technologies—the East River and other urban waterways are unexplored wilderness.

3. Buildings+

For years now, architects have discussed and designed projects based on the premise that buildings no longer require load-bearing walls. After the development of steel and reinforced concrete structural systems, exterior walls — or building envelopes — are rarely called on to provide support. This allows envelopes to play other roles in architecture, such as offering environmental and aesthetic effects. The building envelope, it seems, is now free to do anything the architect wants it to do.

When Peter Cook, one of the founders of Archigram, aimed his great imagination at a new art museum for Graz, Austria in 2000, he pictured a bulbous blue envelope that changed its appearance by glowing and blinking. Cook and his collaborator Colin Fournier observed that the museum had no permanent collection, and they decided that because its contents were meant to change constantly, so should its envelope.

Cook and Fournier—along with consultants Realities:United—developed a multi-layered envelope, including an electronic layer comprised of about 900 "big pixels" named BIX. Each pixel was a simple circular fluorescent bulb about one foot in diameter, and together the pixels formed a low-resolution screen in deliberate contrast to typical high-resolution digital billboards used for advertising. BIX was not like the envelopes of Times Square. It did not render realistic video and it did not have a life of its own independent of the building that spawned it. Instead, the BIX envelope was closely connected to the program and the flows of the building. Its screen reflected on the exterior what was happening on the interior. It surpassed the building envelope's traditional role as a solid, impermeable boundary meant to stop flows of heat, water, and light. BIX provided these functions, but in addition it acted as a permeable threshold and established a cycle of exchange.

Since the opening of Kunsthauz Graz in 2003, dozens of electronic envelopes have come online in cities like Seoul, Shenzhen, and Singapore. All of these envelopes involve low-energy illumination and a pixel-based approach to lighting that can display patterns, graphics, and even text. The envelopes differ in degree of resolution, density of lighting, and content displayed, but they share in common the use of artificial light as a membrane in the city that interacts with multiple flows.

4. Humans+

For the first time in history, more people live in cities than in rural areas. By 2050, 70% of the world will live in large cities. In New York — already the densest city in the United States — the population is growing by about a million people every 15 years. And New York continues to be the most popular entry point for legal immigrants to the United States.

These immense flows of people to and between cities are directly connected to the Carbon Cycle and flows of carbon. Increases in density of humans generally lead to decreases in human carbon emissions. On average, people in New York generate one-third as much carbon as other people in the United States, primarily due to yet another layer of flows—the efficient flows of energy. Dense cities conserve energy through public transportation and through the heating and cooling of small, stacked, party-wall residences.

But clearly the flow of humans to cities cannot by itself ensure a sustainable flow of carbon. All of the flows of the city—including flows of carbon, air, water, buildings, humans, energy, vegetation, and non-humans such as animals and fish—are influential and interdependent. There might be clear relationships between some of these flows, but there is no easy and complete formula for all of them. Some flows we might regulate, but others might be beyond our control. And this only intensifies the importance of the calibrating the city's envelopes, which interact with these flows, and which are entirely at our human disposal.

5. Envelopes=

The city, then, when considered as a site for contemporary design, in the context of complex life-or-death flows, might call for a re-thinking of envelopes. Here are some tentative proposals.

The envelopes of the city could consist of both building envelopes and non-building envelopes. In addition to building skins, they could include tangible planes like streets and the surface of rivers, and intangible planes like neighborhood lines and microclimate boundaries. Just as collaborative teams of architects, engineers, artists, material scientists, and theorists are working together to design specific single building envelopes (as well as new systems and theories of building envelopes in general), these same teams might collaborate on the design of important non-building envelopes.

The envelopes of the city could each be a small ecosystem. As computing becomes smaller, cheaper, and more embedded in everyday objects, envelopes themselves could become complex systems of inputs and outputs. They could collect data from multiple sensors; filter, process, and store information; and respond to changing conditions through morphing, illuminating, and altering other physical properties. In this sense, each envelope might involve its own internal flows.

The envelopes of the city could be networked together. Since computing is becoming more connected (in addition to becoming more embedded), envelopes could take advantage of the capacity to communicate with one another. The building skin could become aware of the river surface. The two could exchange sensor data. Both could have phone numbers so that citizens could text message them and browse the information they collect. The combined envelopes of the city could establish their own flows of data.

The envelopes of the city could be public. Like our parks and like the air we breathe, all envelopes could be considered to be public resources and public spaces. Even building envelopes could be defined in this way. While a building is private and belongs to an owner, the building skin might belong to the life of the street, to the city, and to all citizens. All envelopes—on buildings or elsewhere—might become registers of collective interest and sites for participation. They might address our shared resources and our common concerns. Combine this with Adam Greenfield's inspiring call to action: every public object should have an open API. In other words, assuming that park benches, street lamps, and even streets themselves will soon have embedded sensors and networking electronics, these objects could be openly readable and writable. Since the objects are made public, their data could be made public.

The envelopes of the city could be interfaces to information. Based on their capacity to gather, process, and exchange information, envelopes might offer a wide variety of interfaces to the life of the city. They could display environmental quality, energy use, and public interest in important civic issues. They could communicate about other neighborhoods and other cities. In an era when there is more sensor data being collected than there is hard drive space to store it, the envelopes of the city might play an essential role in real-time filtering and display of information from a variety of sources. This information might be communicated through light or movement or other changes. It might offer direct or ambient information.

The envelopes of the city could go beyond raising awareness—they could engage and solve problems. While envelopes could play an important role in providing information to the public, they might also be considered sites for action. A building envelope might play an active role in a reducing a building's energy consumption by altering its own shading properties and triggering the heating and cooling system based on information about temperature, sunlight, and how much energy other buildings are using. The surface of the river might play an active role in balancing the dissolved oxygen level of the water by pumping in air based on local sensor data and information about the migration of fish and the flows of algae.

The envelopes of the city could be more than hard boundaries—they could be porous thresholds. Envelopes could be sites of negotiation and exchange. They could selectively transfer moisture, nutrients, and electricity, like human skin or cell membranes. Each of the flows of the city might be channeled, filtered, and accelerated in different ways by different envelopes. In this expanded definition, envelopes might not be simply enclosures with clear interiors and exteriors. Instead, they might be a field of inflection points with multiple sophisticated roles.

6. Amphibious Architecture

Early in the fall of 2009, for the exhibition Toward the Sentient City, we installed two networks of sensors and LEDs in the East River and the Bronx River in New York City. One way to understand the project is as a horizontal envelope. It might be considered an experimental building envelope turned on its side and floated out into an underused public space of the city.

Among other precedents, the project built on examples of dynamic building envelopes with embedded, networked electronics. It directly monitored many of the city's flows — including the presence of fish via sonar sensors, the Carbon Cycle and the health of the river via dissolved oxygen sensors, the hydrodynamic flows of the rivers via swaying tubes, and the degree of human attention to the river via text messaging. And it glowed and blinked as a low-resolution display of changing conditions in the water and on land.

This envelope was very small compared to the overall flows of the city. It was modest compared to our proposals for the envelopes of the city. Our horizontal envelope was not yet networked to other envelopes. It did not yet send sensor data to other buildings or cause other envelopes to blink in Times Square or Shanghai. It was still under the radar of most citizens. It may have raised awareness about the rivers, but it did not yet directly remediate them.

But the installation was an essential test. Our small prototype was necessary — not as a demonstration of what we already knew, but rather as an exploration of what we did not yet know. Only through floating the tubes out in the water and equipping them with a text messaging system could we learn about how the public would interact with them (on average, more than three times). Only through marking a public space on the surface of the river could we see if citizens' perceptions of the water and the city's building envelopes would shift (the verdict is still out). Only through blinking our LED disks a foot above the river's surface could we learn if the light would attract or repel fish (the numbers of fish remained the same before and during the installation). Our installation was an open-ended experiment — in the spirit of experiments by Humphry Davy, Verdant Power, Peter Cook, Colin Fournier, and Realities:United — and it was an important step in advancing our research about envelopes and flows in the city.

Amphibious Architecture, installation on East River, 2009.

Our proposals require our tests, and vice versa. Together, the two have their own flows

and cycles. They feed one another. They feed us as architects. And, over time, we hope they might feed the sentient city.

NOTES

1. Richard Holmes, The Age of Wonder (Vintage Books, 2008), p. 454.
2. Shawn Tully, "Water, Water Everywhere" (Fortune Magazine, May 15, 2000).

ENGINEERING AND THE UNDERFUNDED AD HOC PROJECT
The REAL Principles
Jonathan Laventhol and Natalie Jeremijenko

The Amphibious Architecture project placed floating interactive light displays in the rivers of New York City. In engineering terms, the displays were a very coarse array of controllable multicolor LEDs on buoys: each buoy had a ring of red, green, and blue LEDs and each color was independently controllable. The array in the East River had 16 buoys arranged in a 10m by 10m square, the Bronx River display was only eight buoys. Dissolved oxygen sensors in the water give an indication of water quality, ultrasonic fish finders give an indication of the presence of fish, and an SMS interface gives an indication of the presence of humans. These sensors affect the display on the water and feed a database for remote monitoring. Separately, a website conveyed information about the project, but fundamentally the interaction was intended for a local audience human, piscine, avian, and one or two beavers and turtles.

Our project team was geographically dispersed with most distributed throughout New York and one in London. Project communication was achieved predominantly through a wiki, hosted by New Your University (NYU) using WikiMedia software. Out of sheer laziness we simply had one big single wiki page! When printed, it forms a kind of working manual-cum-diary running to 50 pages. The wiki concept worked well, especially with photographs.

Diagrams are absolutely essential for this kind of work. As most of the team was new to electronics and physical computing, we decided to adopt informal approaches to circuit and other diagrams. The most successful techniques were simply to draw on a piece of paper and then use a webcam to grab the image and post it to the wiki or display via videochat.

An early decision was to use the open source Arduino platform for physical computing, which proved to be excellent. As an aside, with equal success one of the authors was also working on a strictly commercial project and used the same Arduino platform to control model racing cars, but with custom printed circuits to implement the Arduino circuit. The principal value of the Arduino platform was the "wayfinding" value of a known approach: we all knew where to look for help, and the core libraries covered almost everything we needed.

The overall system consisted of a number of things connected over the public Internet:
— A back-of-house server
— A text-message input system
— Several Sites

The back-of-house server was a virtual host running a database, a web server, and an extremely simple UDP listening server. The text-messaging input system was commercial and was configured to simply send appropriate HTTP 'GET' instructions depending on the SMS interactions.

Each Site consisted of a number of things connected over a local Ethernet and exchanging messages over UDP:
— A shore-side Macintosh computer
— An input system of sensors
— A display of buoys

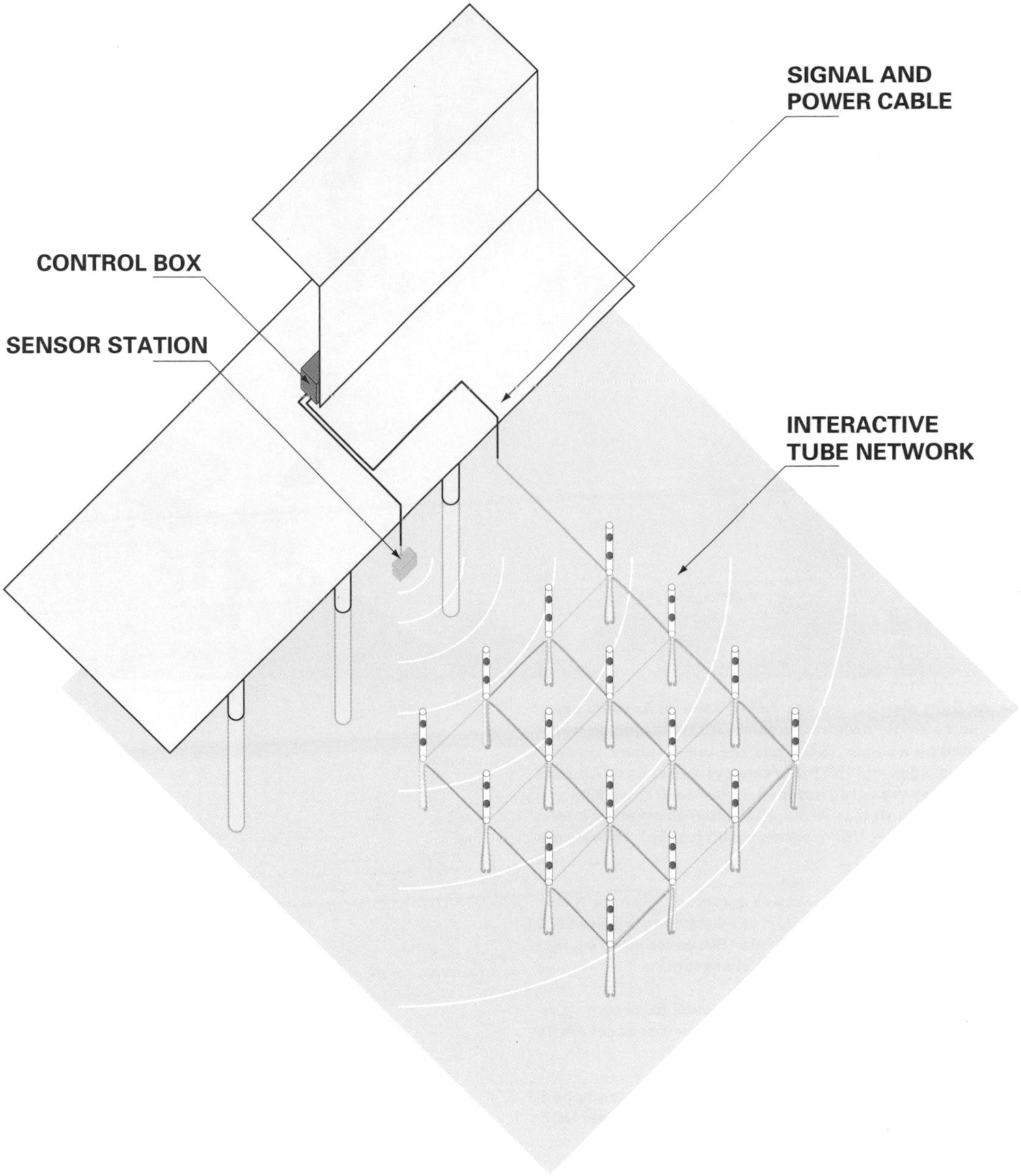
SIGNAL AND
POWER CABLE
CONTROL BOX
SENSOR STATION
INTERACTIVE
TUBE NETWORK

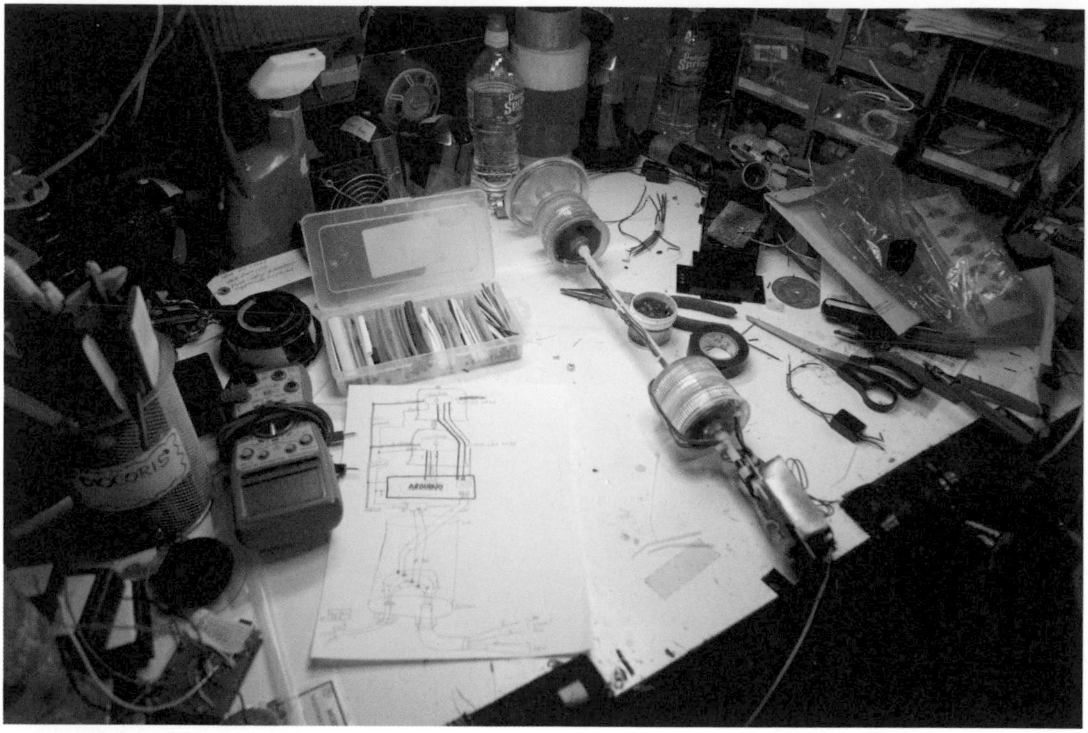

For the light display, we decided to have an Arduino per buoy, and a single Arduino on shore. All the Arduinos were connected on a single RS485 bus; the one on shore always sends messages and the others always listen. Each Arudino in the display was programmed (in hardware) with its address. The shoreside Arduino also had an ethernet interface and listens for UDP packets which it simply repeats out of the RS485.

Remembering that our chain was quite long — about 100m — we decided to run the system at a very conservative 9600 baud. The messages were a simple URL-encoded string for ease of debugging, with a simple checksum.

Each display Arduino waited for a message to its own address, and if found, replayed one of a number of set display sequences which were nicknamed "dances".

In the other direction, the sensors fed an Arduino equipped with an Ethernet interface. This sent out appropriate UDP packets to the shore-side Macintosh computer.

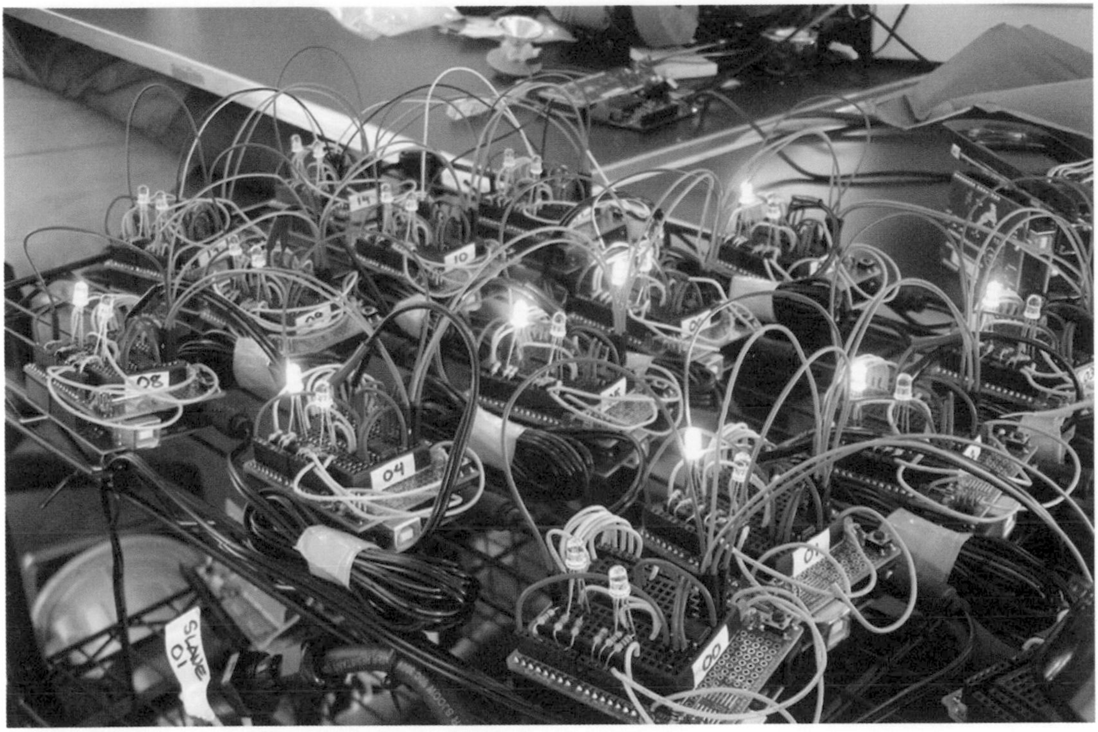

Two of the decisions were key: using UDP and using RS485.

The User Datagram Protocol (UDP) is a key protocol on top of the Internet protocol, and is designed for the case where communicating devices have to send simple messages to each other. In our case, we wanted to be able to instruct the display to perform a certain action. Unlike its more widely-used alternative TCP, UDP is extremely raw. In principle UDP packets can be lost, or even duplicated or arrive in peculiar orders, but in practice these events are rare. The particular property that was of use to us is that UDP provides no indication to the sender whether the packet arrived or not. While this is usually a disadvantage, in our case it means that the sender has no idea at all about the receiver — not even whether the receiver exists — and thus means that the receiver cannot affect the sender in any way at all. This decoupling means that we can test our sender without having built any receivers yet and using simple debugging techniques to determine the proper functioning of the sender. Similarly, because the sending process is so simple, we can send packets from simple command-line programs to develop our receivers in the absence of any finished sender software systems. This engineering decoupling permits project decoupling.

RS485 is extensively used in engineering as the electrical specification for many kinds of long-haul and low-speed communications. In art contexts it is relatively unknown but is extremely easy to use, especially in the configuration we used, where one system is always the speaker and multiple others are always listening. It's the perfect way to build low-cost high-reliability communications at low speed.

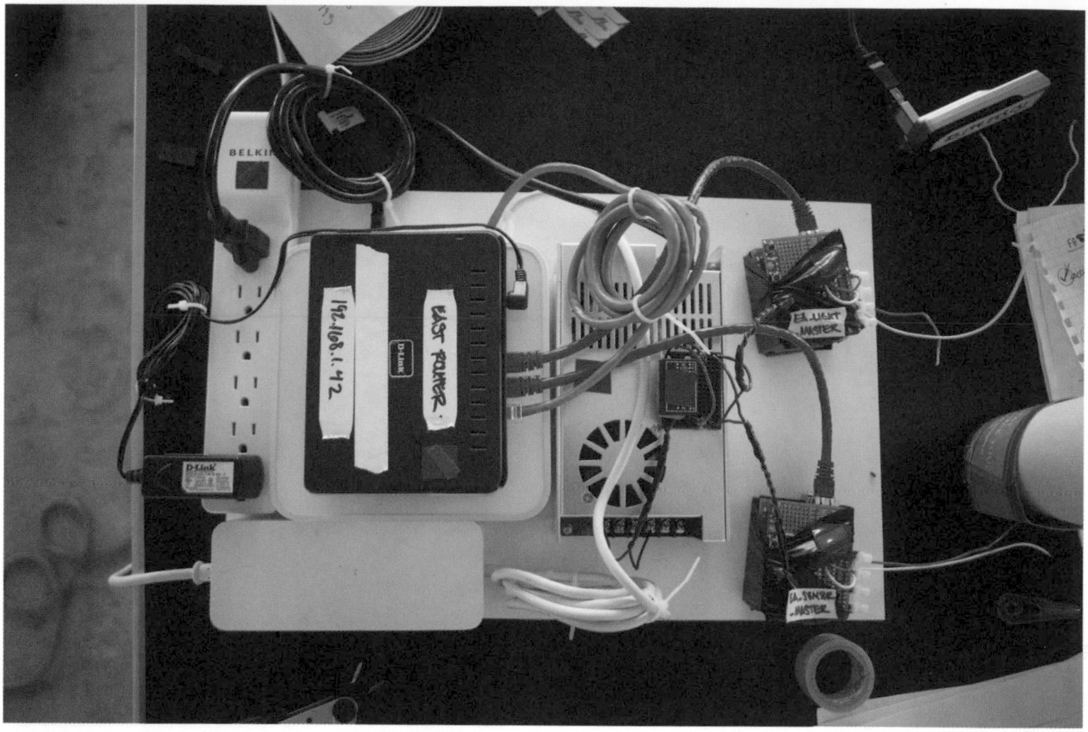

ASSESSMENT OF SYSTEM AGAINST “REAL” CRITERIA

We are kidding ourselves if we think that we are ever going to work on a project with enough funding. No matter how large the budget, somehow the goals manage to be just a little bit larger. Much has been written about project management for large projects such as bridges and software development, and also well-defined projects like films; we are addressing the smaller kind of ad-hoc project, especially those for non-profit organizations. We are thinking of art installations, museum projects, interactive exhibits, festivals and so on.

Often, we are working with inexperienced people who are keen and involved in the project to learn. Sometimes we have amateurs or others who are really in the project for social reasons. Usually, there are some professionals driving the project, and who have very high aspirations for the project, but those professionals are usually stealing time from their day jobs. Typically, there are funders and other gatekeepers who are sceptical and conservative. Typically, no one is getting paid even expenses, and many are funding the project with their equipment, resources and materials as well as their time. The organization can bear more resemblance to a newly-formed punk band than an engineering unit.

We need to accept these aspects and learn to work with them: we are not interested in those who know how to achieve things by hiring professionals, we are interested in those who have decided to achieve something and, hook or crook, they will learn what they have to do to make it happen.

This assesment describes the principles of working under these conditions, which we call the REAL principles: Robustness, Economy, Aspiration, Longevity.

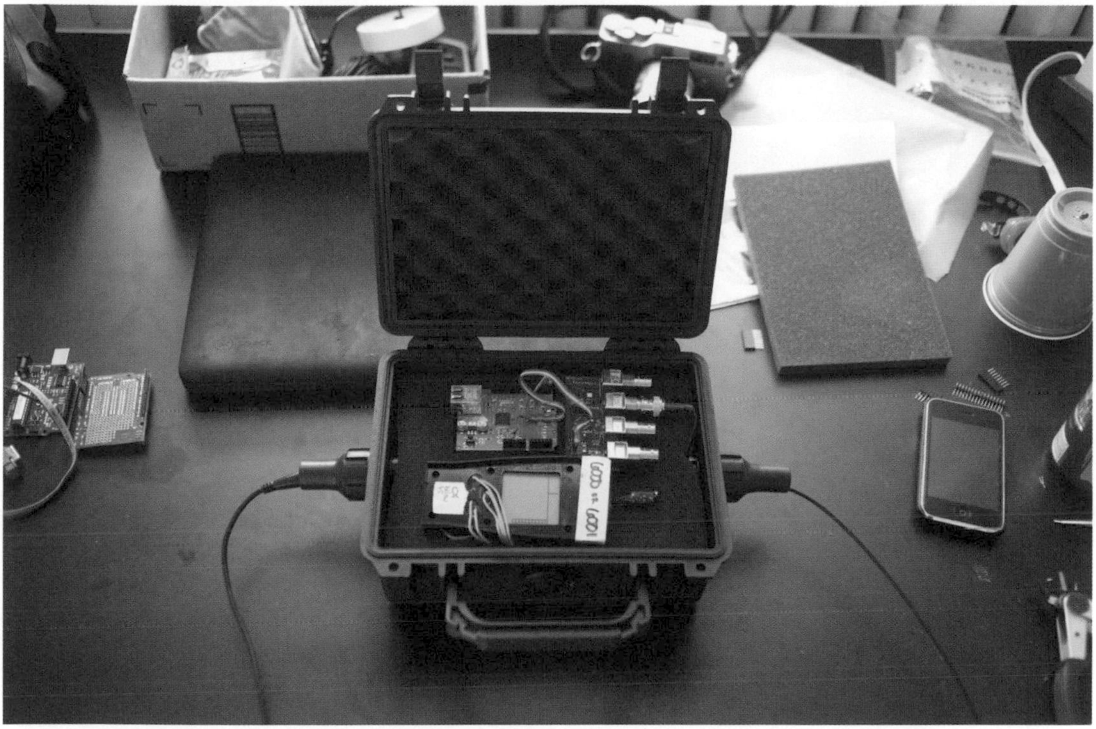

Our project must have
— Robustness: The thing we make has to stand up to its installation; if it falls over or crashes or blows fuses then we have simply failed
— Economy: if we run out of money we will not complete the project
— Aspiration: the goals of the project are typically educational, or conversation-generating, or audience-pulling or similar objectives in mental or social spheres, and we must at least partially achieve these
— Longevity: installations must survive their term

Robust
The decisions on UDP and RS485 were directly driven by the requirement that the systems are robust against noise and the ability to test without our systems being yet finished. The fact that the system ends up in the water meant that repairs would be extremely difficult if not impossible. The decision to use a bus for communication, a single cable for data and power, and have exactly two cables per buoy was specifically designed to minimise water ingress.

Economic
Arduinos are cheap, and building our own RS485 transceivers saved a lot of money while still providing the possibility of interfacing to off-the-shelf parts. We used a single small Macintosh computer on site, and a virtual host in the cloud for back of house. Open source software saved a lot of money here.

Aspiration
The goal of flashing systems in the water was successful. In terms of our meta goals, enormous amounts of conversation were generated.

Longevity
In the end, we were certain that Nature would eat our system. Some of the tides in the East River are surprisingly strong. Nevertheless, our goal of surviving 8 weeks was achieved.

STRUCTURING PARTICIPATION FOR AN ENERGY COMMONS

NATURAL FUSE

HAQUE DESIGN + RESEARCH

USMAN HAQUE

NITIPAK "DOT" SAMSEN

AI HASEGAWA

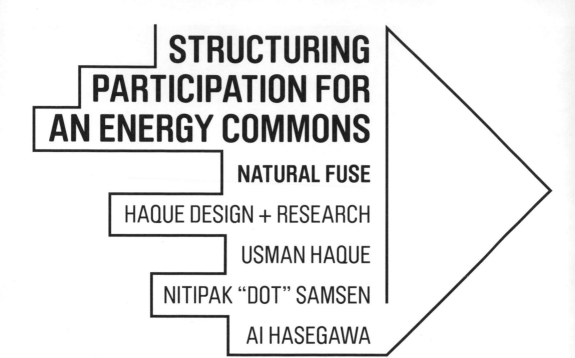

I. Introduction

"Natural Fuse" is a micro scale carbon dioxide overload protection framework that works locally and globally, harnessing the carbon-sinking capabilities of plants. Generating electricity to power the electronic products that populate our lives has an effect on the amount of carbon dioxide present in the atmosphere, which in turn has detrimental environmental consequences. The "carbon footprint" of the power used to run these devices can be offset by the natural carbon-capturing processes that occur as plants absorb carbon dioxide and grow. "Natural Fuse" units take advantage of these phenomena.

Each "Natural Fuse" unit (they are now distributed among households in London, New York and San Sebastian) consists of a houseplant and a power socket. The amount of power available to the socket is limited by the capacity for the plant to offset the carbon footprint of the energy expended: if the appliance plugged in draws so much power that it requires more carbon-offsetting than available, then the unit will not power up.

The problem is that even low-power light bulbs draw more power than can be comfortably offset by a single plant. As a result, all the units are connected together via the Internet so that they can communicate and determine how much excess capacity carbon-offsetting is available within the community of units as a whole.

For example, if you use an appliance that draws 4 watts, and there are six "Natural Fuse" units out in the community that are not currently drawing power, then you can offset the carbon footprint of your appliance by borrowing from others. (Calculations include the energy cost of powering the electronics inside the unit itself too, of course).

The project is as much about the structures of participation as it is about energy conservation. Rather than just having an "on/off" switch for your appliance, you are provided with a "selfless/selfish" switch. If you choose "selfless," then the unit will provide only enough power that won't harm the community carbon footprint. But, if the carbon sequestering capacity of the community is low, the electricity will switch off after a few seconds — though it may be long enough for what you need to do.

If, on the other hand, you absolutely must have electricity (e.g. you hear an intruder in your apartment and you *must* switch on your light at full power), then you might want to choose "selfish" — which will give you as much power as your appliance needs. BUT, if you harm the community's carbon footprint (i.e. it goes from negative to positive) then the "Natural Fuse" system will KILL SOMEBODY ELSE'S PLANT!

Each unit actually has 3 'lives' to lose, before which a vinegar shot is dispensed to the unlucky plant. Thus, as it loses each 'life' an email is sent to both the owner of the dead plant and the owner that sent a 'kill' signal; this provides the capability to communicate and explain their situations to each other prior to final execution of the plant.

Decisions whether to be selfish or not have a visceral impact on others in the community. By networking "Natural Fuses" together, people share their capacity and take advantage of carbon-sinking-surplus in the system since not all "Natural Fuses" will be in use at any one time. If people cooperate on energy expenditure, then the plants will thrive (and everyone may use more energy); but if they don't, then the network begins to kill plants, thus diminishing the network's electrical capacity.

This case study, presented as a collection of consecutive notes and observations, summarizes research on plants, and the development and design of the "Natural Fuse" system, as carried out by Haque Design+Research from January to August 2009. It summarizes our findings, outlines paths followed and describes problems encountered during the process.

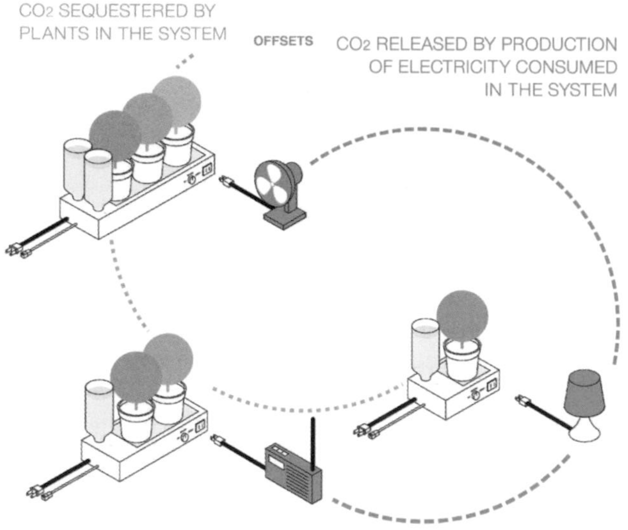

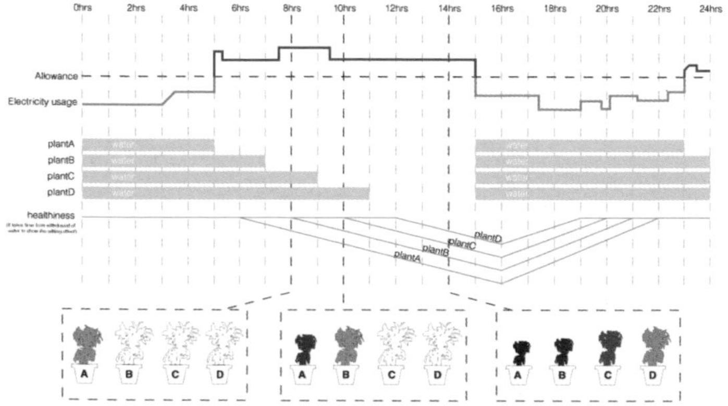

The use of 4 plants as a fine-grained interface is proposed. By staggering the withdrawal of water from the plants, more fine-grained indicator of the electricity network's 'health' is created.

I. PUBLIC EXHIBITIONS

I.I "Natural Fuse" at Futuresonic, Cube Gallery, Manchester, 2009

In May 2009, our preview of "Natural Fuse" was exhibited at the Futuresonic exhibition in Manchester. At this time, we showed some of the developments, problems, ideas and many issues that we encountered.

The prototype: These 4 plants are a part of the network consisting of 10 plants. The system compares the total amount of CO_2 absorbed by plants to the total amount CO_2 emitted by electricity and shows the allowance in a battery-like concept.

The lamp uses the allowance when turned on. If the viewer leaves the lamp on, it flashes and when a sufficient CO_2 allowance has accumulated, the light becomes bright.

The exhibition was divided into 3 parts: the concept, research and problems, and, finally, the goal.

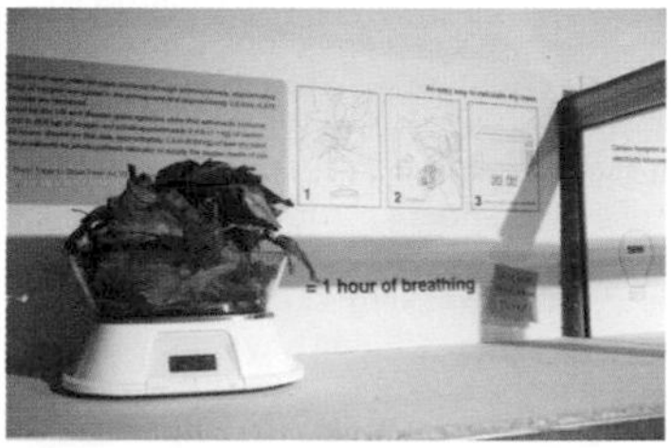

This dry mass contains the same amount of CO2 that a human being releases in 1 hour of breathing.

If a 50W lightbulb is left on for 1 hour, the amount of generated CO2 will be the same as that which is stored in the dry mass and the timber cube.

As a 'network status indicator', we successfully tested the wilting effect of Leopard Lily. It was observed that either a 'dead' or 'alive' effect was achieved, but nothing in between. In this experiment, Leopard Lilies are given water in a 2-hour loop in an attempt to create the effect of 80% or 30% wilting of foliage.

I.II "Natural Fuse" at Toward the Sentient City, 2009

Here is our shop at Toward the Sentient City, 2009. Participants could 'rent' a "Natural Fuse" either by lending plants to the exhibition, or by paying money ($50), which was donated in the participant's name to the Bronx River Art Center.

Participants took a "Natural Fuse" home for 6 weeks. They needed an Internet connection, and a router with an Ethernet port. They took a lamp, a radio or a fan to plug into it.

If the participant 'rented' a "Natural Fuse" by lending plants, then these plants were applied to offset the carbon footprint of making a single cup of coffee. Depending on how many plants were in the exhibition, more cups of coffee could be made! Participants were encouraged to collect their plants at the end of the exhibition.

On the unit, there was a power-activation switch, which the owners could adjust depending on how much energy they wanted to use. There were 3 modes, OFF, SELFLESS and SELFISH.

In "OFF" mode, the system used minimal energy, turning itself on once every hour. No energy flowed to the appliance connected to the unit. As a result, the overall CO2 absorbed in the whole system gradually increased, and the plants in this unit were cared for by the system.

In "SELFLESS" mode, the unit gave power to the appliance at a rate that ensured that CO2 production and capturing in the entire "Natural Fuse" system remained in equilibrium. As a result, the owner was able to turn on the lamp for 10 minutes a day, depending on the status of the whole system and the consumption rate of the appliance.

In "SELFISH" mode, the unit gave as much power to the appliance as it needed. As a result, the owner could use the appliance as usual, but it might cause the total amount of CO2 absorbed in the system to decrease or even lead to total systematic breakdown.

IN CASE A SYSTEMATIC BREAKDOWN OCCURRED, THE SYSTEM WOULD KILL 1 RANDOM UNIT'S PLANTS, AND IT PROBABLY WOULDN'T HAVE BEEN THE OFFENDING PARTICIPANT'S!

Eventually only a few units were rented, and the interaction that we wanted to happen hasn't happened yet. For us, something was missing from this exhibition.

Natural Fuse shop at Toward the Sentient City, 2009.

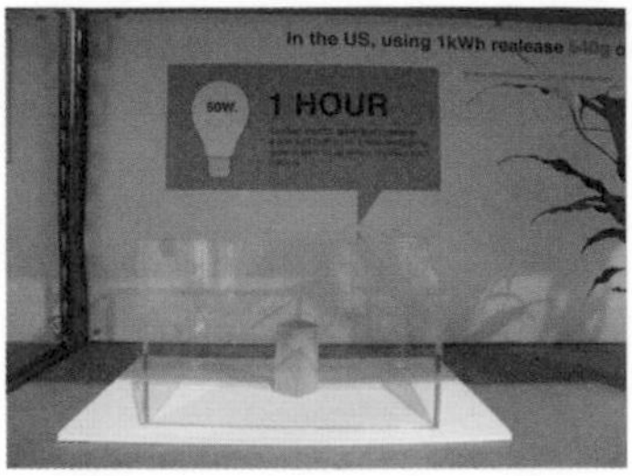

This dry mass contains the same amount of CO_2 as a human being releases in 1 hour of breathing.

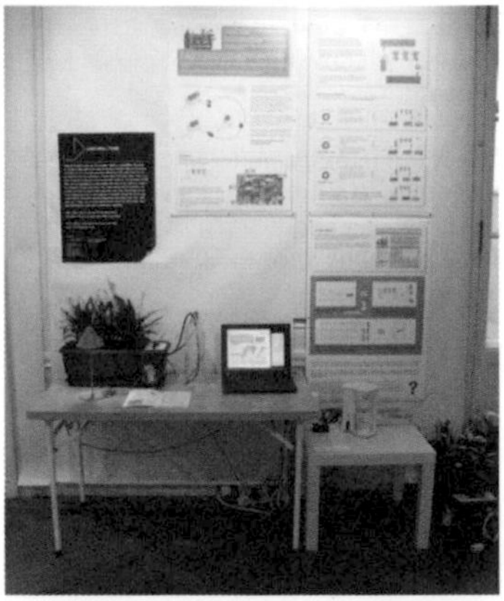

In order to use a 50W light bulb continuously, 380 of these plants are needed to absorb the CO2 emitted.

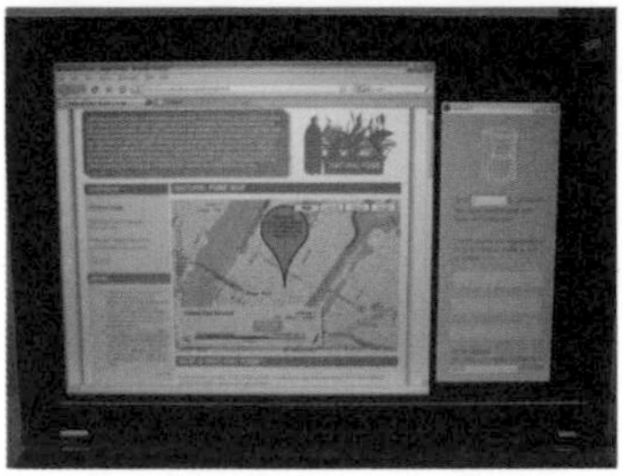

System status and Coffee machine interface.

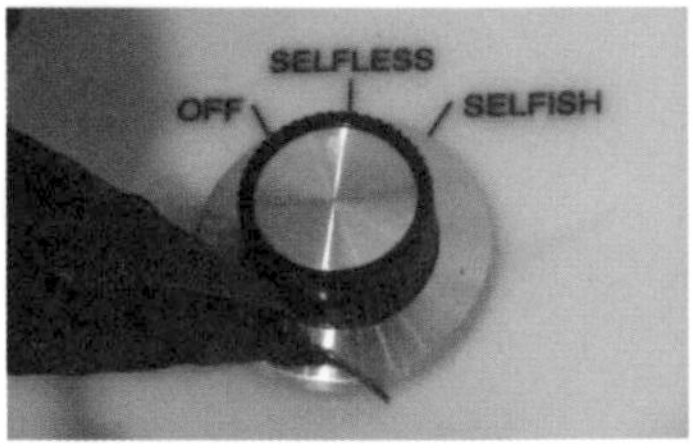

Natural Fuse unit.

I.III "Natural Fuse" at Silicon Dreams, 2010

In February 2010, "Natural Fuse" was exhibited at Silicon Dreams, San Sebastian, Spain. In this version, the project was developed to create a more 'rental shop-like' atmosphere that we think was missing in New York.

This exhibition turned out to be quite successful from our perspective, with all units rented in the first two weeks. In the third week of the exhibition, there was a huge mass plant murder event. Every unit that was online at that time was killed. We tracked down who the killers were: three units that we left in the exhibition as demo units had been left in "Selfish" mode, consequently killing all other online units. An ironic tragic ending...

Natural Fuse shop at Silicon Dreams, 2010.

II. CARBON SEQUESTERING: RESEARCH AND STRATEGY

During research and development, several issues were encountered that affected the design process. These highlight the range of challenges faced by "carbon sinking" initiatives in general. For example, the amount of CO_2 that a single houseplant can sink is much smaller than expected, raising the question as to what to do: Use less energy? Or, greatly increase the size of the fuse? The latter could lead to a need for 402 plants to offset a 50W light bulb. Then, when a plant dies any carbon sequestered during the growth period is, in the absence of continued sequestration (e.g. by sealing it deep within the earth), soon released back into the atmosphere. A zero-sum situation depends entirely on where the arbitrary boundaries of the system are drawn. How should the plant be disposed of? Should it be eaten? Buried? Woven?

1. Carbon sink calculation

First, the question of whether or not a single plant is capable of offsetting the energy expended by a light bulb was examined. In order to define this, the actual energy calculations used to determine carbon offsetting were examined and a way was found to calculate the CO_2 sink capacity of a plant by measuring a plant's dry mass. In this first experiment, the CO_2 sink capacity of the plant was calculated.

(There is an easy, practical and accurate way to do so, as described in "How to Grow Fresh Air, 50 houseplants that purify your home or office" by Dr. B.C. Wolverton.)

On the basis of a more precise set of calculations, the following estimates were made:

"For every dry ton of new plant biomass produced through photosynthesis, approximately 1.4 tons (1,273 kg) of oxygen are added to the atmosphere and approximately 1.8 tons (1,273 kg) of carbon dioxide are removed. Studies conducted by the US and Russian space agencies show that astronauts consume approximately 2.0 lb (0.9 kg) of oxygen and exhale approximately 2.4 lb (1.1 kg) of carbon dioxide every 24 hours. Based on this data, approximately 1.5 lb (0.64 kg) of new dry plant material must be produced by photosynthesis each day to supply the oxygen needs of one adult." [1]

In other words, the amount of CO_2 that an average-sized houseplant is able to absorb per day is so little that an immense number of plants would be needed to balance one human's daily CO_2 production. In this situation, what had to be determined was how to deal with this conceptually. One of the possibilities was to make a symbolic gesture. In the "Natural Fuse" project, a symbolic gesture and the psychological impact were considered. However, there is a way to strengthen the impact.

The table at right reveals the sources of energy in the USA.Calculations reveal the following:

— The average carbon footprint per energy usage in the USA - 322g of CO2/1 kWh*

— Given the use of a 50 watt light bulb = 0.05 kWh = 16.1 g / hour.

— Each of these plants can sink 0.99 g of CO_2 a day = 0.04 g / hour.

— Therefore 16.1 / 0.04 = 402 of these plants are needed to sink the CO_2 instantly.

— On the other hand, if 1 plant is used, it can sink 1 g / day.

— This allows for the use of a 50 w light bulb for 3.7 minutes a day.

So it may be possible to manage how energy is used over time more efficiently, or use "Natural Fuse" to

Left to right, top to bottom: The wet mass of the leaves is 181 g., the dry mass of the leaves is 25 g., the wet mass of the root is 139 g., the dry mass of the root is 8 g.

1 ton of plant dry mass ▸▸▸ 1.8 tons of CO_2 removed from air
33 grams of plant dry mass ▸▸▸ 59.4 grams of CO_2 removed from air

1 person exhales approximately 1.1 kg of CO_2 each day

If this plant takes 2 months to grow to this size,
each day this plant removes 0.99 gram of CO_2

Then we need 1,111 of these plants to cover our 1 day exhaled CO2 !!!

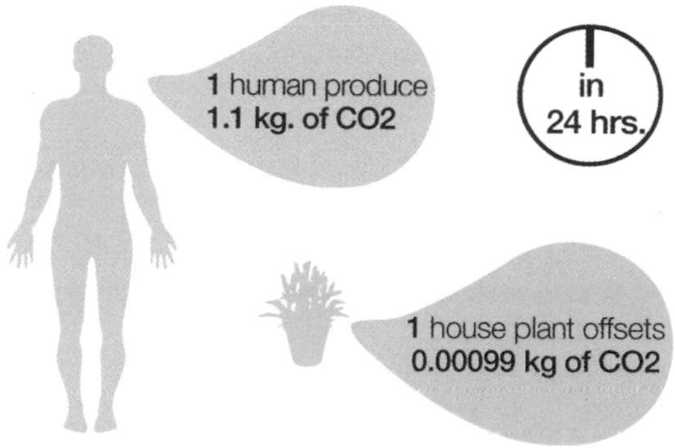

1 human produce
1.1 kg. of CO2
in
24 hrs.
1 house plant offsets
0.00099 kg of CO2

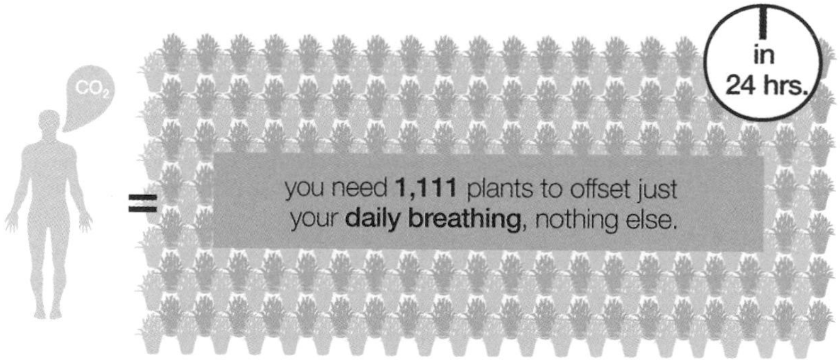

CO_2
=
in
24 hrs.
you need **1,111** plants to offset just
your **daily breathing**, nothing else.

source in the usa	percent	CO_2g/kWh	
coal	48.9	450	24450
petroleum	1.6	650	1040
natural gas	20	330	6600
nuclear	19.3	5	96.5
hydroelelectric	7.1	5	35.5
other renewable	2.4	5	12
other	0.7	5	3.5
			32237.5
average			322
			CO_2g/kWh

suggest amended energy use patterns in people.

2. Two options
On the basis of the information obtained from the experiment with "LeopardLilyB04" the following calculations were made, and an inconvenient truth became evident:

A. With the use of a 50W light bulb, 402 of these plants are needed to sink the CO_2 instantly. Each of these plants will allow for the use of a single 0.125W appliance continuously.
B. On the other hand, each of these plants allows for the use of either a 50W appliance for 3 minutes a day, or 10W for 15 minutes, 5W for 30 minutes or 150W for 1 minute.

The first option is similar to the creation of a conventional fuse, allowing for the use of maximum current per 1 second. If the calculation is accurate, this option is far too demanding to operate successfully. The second option appeared to be more viable, as plants sink CO_2 continuously during daylight. This offers maximum energy per day, somewhat like an accumulated fuse or a half-battery-fuse. For example, the LeopardLilyB04 offers 9000 joules/day. (Joules = energy of 1 Watt for 1 second).

If there are 4 of these Leopard Lilies, it might be possible to use a 10W lamp to read a book for 1 hour before sleeping (and before the circuit breaks).

3. Electricity usage
After two months, when the Leopard Lilies had grown considerably larger, another calculation was made to discover for how much time energy could be used. Accordingly, research indicated that:

— In order to use a 50W light bulb continuously, 380 Leopard Lilies are needed.
— In order to use a 2.5W (1/20 of 50W) light bulb continuously, 19 Leopard Lilies are needed.
—In order to use it only 1 hour per day, 19/24 = 0.79166666667 plants are needed.
— In order to use it only 3 hours per day, (19/24) x 3 = 2.375 plants are needed.
— One Leopard Lily can sink 1.263g of CO_2 per day (24h/19plants = 1.26315789474).
— If there are three Leopard Lilies of the same size, an Ikea 3 Watt LED lamp can be used for 3 hours 47 minutes 34 seconds (1.263g x 3 plants = 3.789).

4. Factors influencing CO_2 absorption
The amount of CO_2 absorbed and released by plants depends on various factors, i.e.:

- CO_2 concentration in the atmosphere
- water available to the plant
- the number of leaves present on the plant (i.e. number of chloroplast)
- the intensity of light
- the region where the plant is grown
- temperatures

Accordingly, it is not possible to find an average amount of CO_2 sequestered and released by plants as this is dependent on many individual factors.

TEMPERATURE: Photosynthesis is most intensive in a temperature range between 20°C and 30°C (68°F and 86°F). Lower and higher temperatures hamper the process.

HUMIDITY: The uptake of carbon dioxide from the surrounding air happens through stomata - microscopic openings on the underside of the leaves. Humidity and temperature ranges trigger the opening and closing of these pores to regulate internal processes. If the air is too dry, photosynthesis slows down. Ideal humidity ranges are 40-60%.

LIGHT: Reductions in sunlight also reduce photosynthesis. However, some plants need more and some less light.

MINERALS: Additionally, minerals are crucial for photosynthesis (and in consequence influence CO_2 absorption). Lack of just one ingredient influences the process as well.

The more CO_2, the better the absorption. A rise in CO_2 levels (as well as in brightness of light and amount of water) increases the speed of photosynthesis: a 0.15% rise of CO_2 concentration causes a triple increase in intensiveness of photosynthesis, i.e. arable plants in greenhouses can "fatten up" with CO_2: in the air we breathe, the concentration of CO_2 equals 300-400ppm [3], and most of the plants require 800-1000ppm for maximum production, i.e. the more they get, the better they grow and the more they absorb. Cacti and succulents absorb CO_2 in a different way because they must survive in a dry environment and are therefore obliged to administer water efficiently by absorbing CO_2 at night.

5. Individual carbon sink capacity of plants
The question arose as to what other plants would be the most efficient in terms of carbon sink capacity and if there were plants suitable for the project with a higher CO_2 absorption ratio than the Leopard Lilly. However, two conditions had to be fulfilled: plants that were both easy to purchase in a big city (e.g. in markets) and also capable of thriving in an indoor environment. These terms disqualify both of the following possibilities:

a. Trees:
The obvious object of interest became trees which, due to their bio-mass, display the highest carbon sink ratio. A tree can absorb about 50lbs of CO_2 per year (depending on its age, species and geographical location). However, trees could not be used for the purposes of the project, because those more readily available (from moderate and continental climatic zones) need cold winters and fairly cool springs and autumns. If a tree is placed indoors in a heated room i.e. in autumn, it will die because it will expect winter to come. Theoretically, it is possible to grow trees indoors, but it is necessary to create specific conditions (periods of cold; rather difficult to achieve for this project). One solution might have been to grow exotic trees (from tropical or dry climate zones, although tropical trees require high humidity, a further challenge). Another argument against using trees was their height, which would require very tall spaces.

b. C4 Plants:
There is a group of plants called "C4 plants" that are the best CO_2 absorbents, such as sugar cane and corn. However, these are difficult to purchase and cultivate indoors in an urban environment.

6. Disposal of dead plants
The plants only act as a "carbon sink" if their remains are disposed of after they die in such a way that the carbon is prevented from escaping into the atmosphere.

a. Consume the plant – CO_2 circulation in the human body

During the process of photosynthesis, plants absorb CO_2 which is transformed into saccharides and then into proteins and fat. When people and animals eat vegetables, for example, they use the substances contained in organic compounds to: (i) build up their bodies and (ii) produce energy. There are two kinds of respiration (both deal with CO_2): (i) breathing - gas exchange, (ii) tissue/cellular respiration - the set of metabolic reactions and processes that take place in an organism's cells to convert biochemical energy from nutrients into ATP, a universal compound carrying chemical energy. Released energy is used to make the body work. In this second process, nutrients - carbohydrates, fat and proteins (and CO_2 contained in the nutrients of e.g. vegetables) are oxidized (transform into CO_2). This CO_2 is then used to build bones, hair, kidneys, etc. An adequate level of CO_2 in human blood determines the oxygenation of tissues and the correct functioning of enzymes and organs.

One liter of blood leaving the lungs contains approximately 500 cm3 of CO_2 (40 mmHg); one liter of blood entering the lungs contains approximately 550 cm3 (46 mmHg). Humans get rid of CO_2 through excretion and defecation (as undigested food materials also contain the compound). Furthermore, during the process of decay of dead substances, CO_2 goes back into circulation (when a human dies, CO_2 contained in the body goes back into the ground or the atmosphere). So, if a plant is eaten, the CO_2 sequestered by the plant eventually ends up back in the atmosphere: either via respiration, via decay or via defecation.[4]

b. Bury the plant underground

The technique of sealing captured CO_2 deep within the ground has already been implemented in several countries. Geological sequestration involves injecting CO_2 into underground rock formations below the Earth's surface. These natural reservoirs have overlying rocks that form a seal, keeping the gas contained. Basalt formations (volcanic rock) also appear to be suitable for storing CO_2. In fact, basalt is one of the most common types of rock in the Earth's crust - even the ocean floor is made of basalt [source: USGS]. Researchers have found that when they inject CO_2 into basalt, it eventually turns into limestone - essentially converting to rock.[5]

There can be risks to underground storage on a big scale such as leakage, thereby increasing the acidity of soil and water. As for the single plant, sealing it underground would be costly and involve a much higher production of CO_2 than the plant itself contains.

c. Conversion to a useful physical object

Another possibility is to transform the plant into something else, such as a physical object with long-term use value. One could, for example, weave plant foliage to produce fabric.

III. EXPERIMENTS

1. PLANTS

An attempt was made to answer the question of how to display the ratio of energy consumption in the system. The greatest challenge was to find and elicit a reaction that would be reversible. Through the research and experimentation, a number of plants were selected that might be used successfully in this project: their reaction to stimuli is reversible and sufficiently fast to be noticeable. The most promising and effective appeared to be an effect of wilting (as a 'network status indicator,' we successfully tested the wilting effect of a Leopard Lily, a Water Plant and a Chrysanthemum), however, due to difficulties in controlling this, the idea was abandoned, ipso facto acknowledging the priority of the plant's carbon sink capacity. Several ways of killing the plants relatively easily were found.

Experiments were preceded by research, which enabled the location and selection of plants that could be used effectively as indicators (i.e. they react to certain stimuli).

There were several elements that needed research and experimentation. First, we needed to select plants that were not too sensitive to the varied and sometimes sudden changes in environmental conditions (i.e. temperature, humidity, level of light available) and, at the same time, that were responsive to certain factors in a visible and distinct way. Therefore, for the purpose of the project it was decided to use a number of houseplants that were neither demanding nor difficult to grow. Since speed of reaction and its reversibility were important, if not crucial, in this case, the aim was to check how easily and quickly they respond to some favorable/unfavorable conditions. We also investigated using hydroponic techniques for plant growth and sustenance, since this method might provide the quickest results, and because there was a higher likelihood that the emergent condition of plants weaned in this manner might be more reversible.

Issues of focus:

1. FACTORS TO WHICH PLANTS RESPOND: lack of water, overwatering, chemicals/organic substances (ammonia, acetic acid, ethylene, calcium, copper, washing powder, bleach, herbicides, fertilizers), dyeing flowers or leaves (food coloring), forced flowering, change in pH levels of the soil, soil salinity, hard water, change in position/rotation of the plant container, touch and electrical currents;

2. PHYSICAL REACTION: change in the color of flowers, change in the color of leaves, fading of leaves, movement, change in taste/smell;

3. REACTION TIME AND RESEARCH ON REVERSIBILITY OF THE EFFECT: the goal was to find the most immediate effect of a change in plant state and find a way to reverse this reaction;

4. PLANTS GROWN WITHOUT SOIL (HYDROPONICS): research and experiments on species that can be grown in this way.

To produce and accomplish the project it was first necessary to carry out a number of experiments. Selected species of plants (several items of the same species) that might be appropriate were purchased and, over a period of several weeks, their reactions to different conditions/doses of substances applied during different time periods was observed. The selection of plants chosen for the experiments also depended on their availability.

Criteria governing which plants would be used in the experiments may be categorized as follows:

1. Lack of water: some plants like Leopard Lily and Hydrangea react rapidly to a lack of water. However, the main issue is reversibility.

2. Over-watering: there may be a longer reaction time than to a lack of water. Examples of plants in this category are Christmas Cacti and Areca Palm.

3. Chemical and substances: multiple tests have shown that some plants react to Ammonia, Acetic Acid, Calcium or Copper. The most common reaction is in a change of color of the flower when reacting to Ammonia.

4. Fertilizer: excessive fertilizing can cause minimal leaf burn in Leopard Lily and can also cause yellowing of foliage in Areca Palm.

5. Color Dyeing of flowers or leaves: many plants and their flowers respond to food coloring, but whether this is reversible was not known as this stage.

6. PH of soil: changing the color of Hydrangea flowers by adding Aluminium Sulfate/Sulphur or lime to the soil is extremely popular, although this process may take three months to complete.

7. Soil salinity/ hard water: leaf color change in many plants, such as Gerbera and Freesea, is a common result of watering with hard water.

8. Orientation of the pot to light: a change in direction of the light source may cause Madagascar Jasmine to lose leaves and flowers.

9. Touch: Mimosa folds its leaves under various stimuli such as touching, warming or shaking.

Experiment: 1.1
Water plant in a container filled with water — aim: to check its reaction to a lack of water (how long it takes for the plant to wilt, whether it can recover and speed of its recovery.)

The container has been refilled with water: the plant recovers, but appears less healthy than before.

8.00 pm 09/02/09: The start of the experiment: water is removed from the water plant's container.

The lid has been removed. After 2 hours the plant wilted.

12/02/09: No change is observed.

Experiment: 1.2
Pansy plant grown in soil watered with vinegar (acetic acid) - aim: to check if the flowers change color under the influence of vinegar.

According to the description of an exercise for school children (text in Polish available here: www.edukacja.warszawa.pl/plik.php?id=2105) the color of pansies changes under the influence of acetic acid and ammonia - blue flowers change to red or pink, red flowers become yellow. It is recommended to use cut flowers, however, whether this change is possible in plants grown in soil still remained to be checked.

It has been observed that it is possible to change the color of flowers by an application of vinegar, which unfortunately ultimately kills the plant.

4.00 pm 10/02/09: The start of the experiment: the pansy B03 has been watered with 25ml of distilled vinegar.

11.00 am 11/02/09: The plant has wilted.

11.00 am 12/02/09: The plant is dead.

Experiment: 1.3
Three Leopard lilies - S01, S02 and S03 - grown in soil. The aim is to check their reaction to a lack of water, to hard water and to an overdose of fertilizer.

A concentrate of dissolved garden lime to turn water and a concentrate of liquid fertilizer.

1.3.1 Leopard lily (S01) grown in soil, watered with a solution of water and lime (hard water) - aim: to check the plant's reaction to hard water.

Leopard Lily S01 watered with hard water: at the start of the experiment.

The plant 1 week later.

1.3.2 Leopard lily (S02) grown in heavily fertilized soil - aim: to check its reaction to an overdose of fertilizer.

Leopard Lily S02 heavily fertilized: at the start of the experiment.

The plant 1 week later.

1.3.3 Leopard lily (S03) grown in soil, deprived of water — aim: to check its reaction to a lack of water.
The plants are resistant — changes were not easily discerned by the end of the week.

Leopard Lily S03: at the start of the experiment.

The plant 1 week later.

Experiment: 1.4
Remove water from 3 water plant containers and:

- retain the lid (B01),
- remove the lid (B02),
- add salt (to absorb moisture; B03)

Aim: to compare the reactions of these three plants (i.e. speed of wilting) and to check if the plants can recover.
A set of experiments were conducted with water plants (B01, B02, B03): water from the three containers was removed. The lids were removed from B02 and B03 and the lid of B01 was retained. Then, salt was added to the container of B03 to remove humidity. It appears that removal of the lid is a more important factor in making the plants wilt than the addition of salt.

Next, water was added to container B02 sufficient only to cover the pot and the lid was replaced. The image below shows the condition of the plants 2 days later.

It was concluded that the reaction time of a water plant was about 1-2 hours after removal of the lid. However, recovery seemed to be a serious issue.

Water plant B01 is still healthy without any addition of water; the lid and a small amount of water can save B02's life, but cannot return it to health. B03 has obviously not recovered and has died.

Experiment: 1.5
Another experiment on water plants (C1, C2 and C3) - this time spraying with water to increase humidity in order to observe whether they can be returned to their initial condition in this way.

1.5.1
It was observed that C2 wilted while C1 and C3 remained in good condition. It was inferred that results due to removal of container lids were inconclusive.

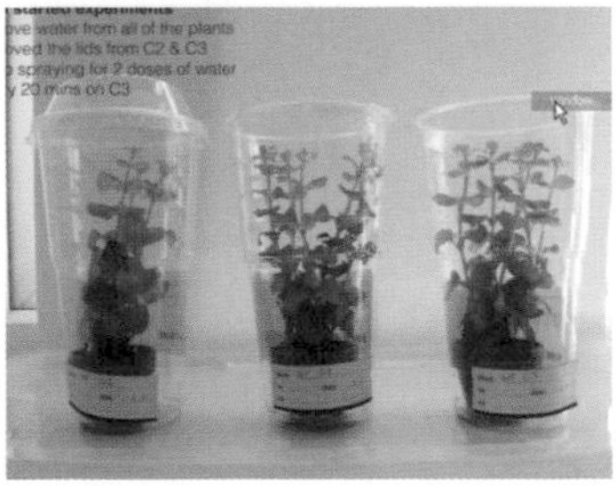

1.54 pm 04/04/09: The start of the experiment: water removed from all of the plant containers; container lids removed from C2 and C3; repeated spraying of C3 with 2 doses of water every 20 minutes.

4.41PM 04/04/09: Started spraying C3 with 2 doses of water every 20 minutes.

7.02PM 04/04/09: End of the experiment.

1.5.2
The result of this experiment is more promising than the previous one: it was observed that over a period of approximately 6 hours the plants returned to their original condition faster than in 1.5.1.

1.12 pm 05/04/09: The start of the experiment: water removed from all containers with plants; lids removed from only C2 and C3.

7.13 pm 05/04/09: Started spraying C2 and C3 with 2 doses of water every 60 min; replaced lids on C2 and C3. It was observed that C2 wilted while C1 and C3 remained in good condition.

After replacing the lids on containers C2 and C3, plant C2 recovered.

Experiment: 1.6

In this experiment, the wilting effect and possibility of color change under the influence of food dyes was examined. In order to do so, the following plants were used:

– Hydroponic Leopard Lily A with roots dipped in a solution of red food dye – aim: to check if the plant changes color under the influence of food dye.
– Celery dipped in food dye - aim: to check if a change in color under the influence of food dye is observed;
– Hydroponic Leopard Lily B deprived of water in its container - aim: to check how fast the plant wilts;

The result:
– Hydroponic Leopard Lily A didn't change its color under the influence of food dye
– Celery's color changed to red started from the top edge to about 2cm below, however its color is irreversible by the blue dye
– Hydroponic Leopard Lily B took about 18 hrs after removing the water to wilt.

The plants used in this experiment.

10.11 am 06/04/09: After 13 hours.

8.41 pm 06/04/09: After 23 hours food dye in the celery has resulted in color changing to blue.

9.25 pm 05/04/09: The start of the experiment.

7.47 pm 06/04/09: After 22 hours.

9.04 am 07/04/09

DELAY PERIOD

From experiments with removing water from both Leopard Lily and water plants, it was seen that there were some patterns in the delay period (as shown in the diagram). For the water plant, a period of 1 hour was observed where there were no changes, followed by a 2 hour period of wilting. When water was poured back into the container, a period of 3 hours was observed where there were no changes followed by a 3 hour period of 'recovery.' For the Leopard Lily this process took much longer, but the effect was greater.

This means that users might not be able to notice anything during a 'delay period'. If the delay period (the time when no changes occur) were moved to the previous indicator period – it may be possible to make the following statement: "the water plant is an indicator of average power consumption in the previous 6-hour period".

1. 00:00 – 06:00
2. 06:01 – 12:00
3. 12:01 – 18:00
4. 18:01 – 00:00

At the start of each period users may be able to see indicator changes.

Experiment: 1.7

Cut pansy flowers placed in a 30% vinegar solution - aim: to check if the flowers change color under the influence of vinegar, since the experiment with the pansy plant grown in soil was unsuccessful, the possibility of changing the color of cut flowers using the same solution is examined.

Result: Pansy didn't change its flower color under the influence of 30% solution of vinegar, but it wilted and eventually died. This experiment should be repeated to check if wilting can be reversed while both flower and stem are wilted but still alive.

9.00 am 09/04/09: The start of the experiment: the flower has been placed into a vinegar solution.

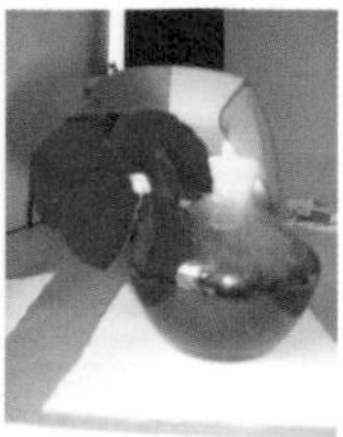

10.50 am 09/04/09: The stem has wilted while the flower itself is still in good condition.

12.10 am 09/04/09: The stem has wilted completely.

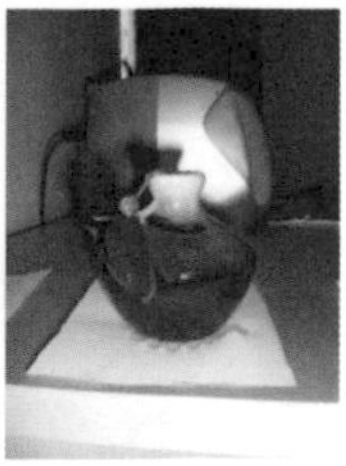

5.00 pm 09/04/09: Both flower and stem have died.

2.00 pm 10/04/09: Both flower and stem are dead.

water plant

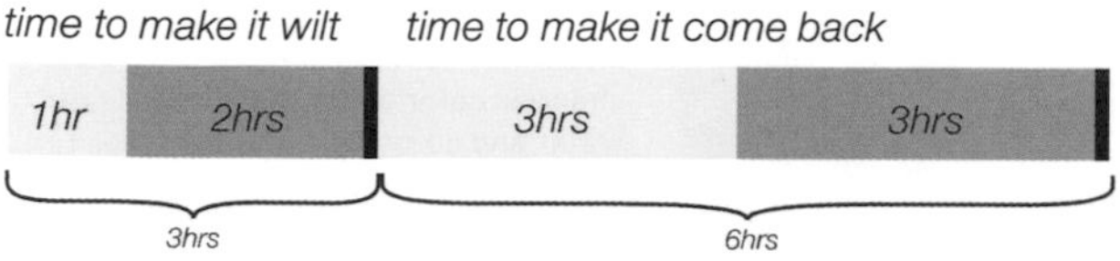

Leopard Lilly

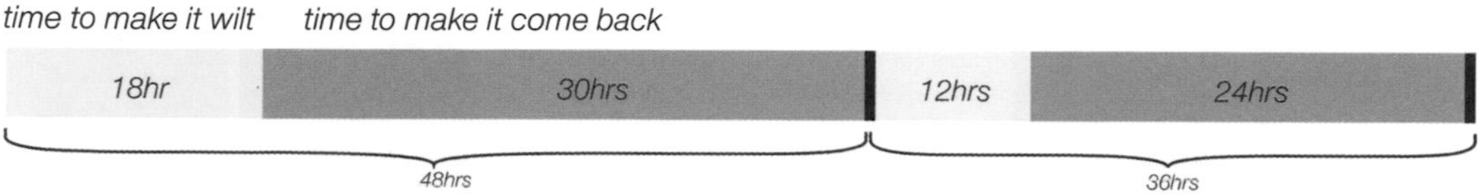

water plant

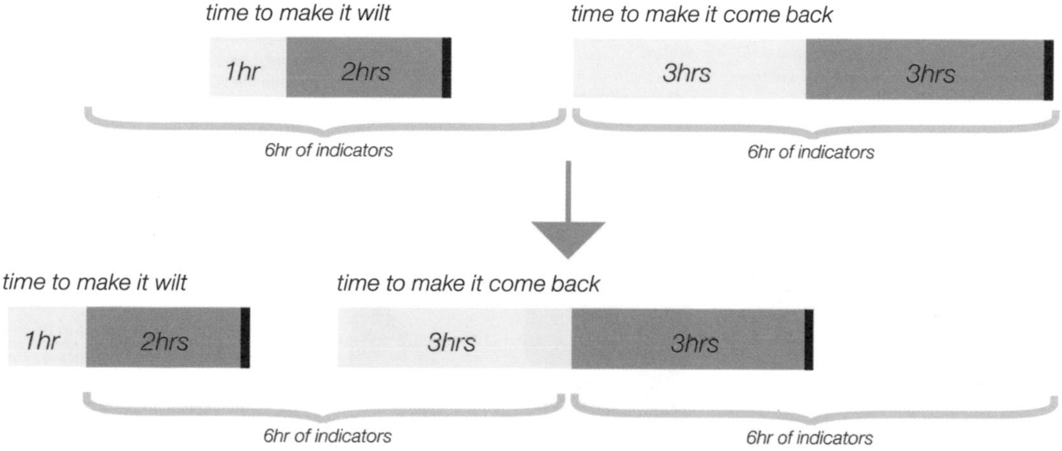

9.15 am 09/04/09: The start of the experiment: roots of the plant have been put into the vinegar solution.

4.40 pm 09/04/09: The flower is dead; the rest of the plant looks well. 5.00 pm 09/04/09: The plant looks quite well, but the flower is dead.

7.30 pm 09/04/09: Lower leaves start to wilt.

6.00 pm 10/04/09: All leaves have wilted completely and become yellowish and an attempt to reverse the effect at this stage is made.
6.30 pm 10/04/09: a solution of vinegar replaced with fresh water.

5.00 pm 09/04/09: Both flower and stem have died.

No change (the pansy plant has not recovered).

Experiment: 1.8
Pansy plant grown hydroponically in a 30% vinegar solution - aim: to check if the flowers change color under the influence of vinegar; to compare reaction times of plants grown hydroponically to plants grown in soil.

Result: the effect of vinegar solution can not be reversed.

Experiment: 1.9
Pansy plant grown in soil watered with a 30% vinegar solution - aim: to check if the flowers change color under the influence of vinegar; to compare reaction time of plants grown hydroponically to plants grown in soil.

Result: Pansy didn't change its flower color, however it took longer to wilt than the Pansy that was grown hydroponically.

9.20 am 09/04/09: Start of the experiment: plant has been watered with 30% vinegar solution.
4.40 pm 09/04/09: No change (the color of the flower has faded, but this is barely noticeable).
7.30 pm 09/04/09: Flower has wilted completely. There has been no change in its leaves, which are still in good condition.

2.00 pm 10/04/09: The stems of the flowers are dead, the rest of the plant appears to be in fairly condition.

10.00 am 11/04/09: Lower leaves have begun to wilt.

Experiment: 1.10
Pansy plant grown hydroponically in water - aim: to check whether pansies can be grown hydroponically. If so, to find out how much time is needed for the plant to wilt after removal of water, whether it can recover and how long this takes.

It was observed that recovery was possible to the detriment of the overall condition of the plant.

9.35 am 09/04/09: Start of experiment: roots of plants are immersed in water.

2.00 pm 10/04/09: No change - the plant looks well.

10.00 am 11/04/09: Water is removed from the jar.
2.00 pm 11/04/09: The whole plant has wilted (both leaves and flower stems).

2.00 pm 11/04/09: The jar is refilled with water.
2.45 pm 11/04/09: The plant has recovered well although its flower buds do not look as healthy as before the experiment).

Experiment: 1.11
Cut pansy flower placed in a container with a solution of yellow dye and water - aim: to check if the flower changes color under the influence of dye and water.

Observations revealed that the overall appearance of the plant gradually worsens during this process.

10.40 am 09/04/09: The start of the experiment: the flower is put into the solution of yellow water to check color change.
4.40 pm 09/04/09: No change.

2.00 pm 10/04/09: The color of the flower has faded slightly, but is barely noticeable; the edges of the petals have become 'rough' and dry. The change is barely noticeable
5.00 pm 10/04/09: Solution of yellow water replaced with water to see if the slight change is reversible.

No improvement - the flower is dying; no signs of recovery.

Experiment: 1.12
Pansy plant grown hydroponically, dipped in a solution of yellow dye and water - aim: to check if the flower changes color under the influence of water and coloring.

Over a period of three days, no change in the plant's appearance was observed.

10.40 am 09/04/09: The start of the experiment: roots of the plant have been put into the solution of yellow water.
4.40 pm 09/04/09: No change.

2.00 pm 10/04/09: No change.
10.00 am 11/04/09: No change.

Experiment: 1.13
Cut pansy flower dipped in a 30% vinegar solution - aim: to check if direct contact with the solution influences the color of the flower.

Changes in color were noted in both the solution and the flower tested.

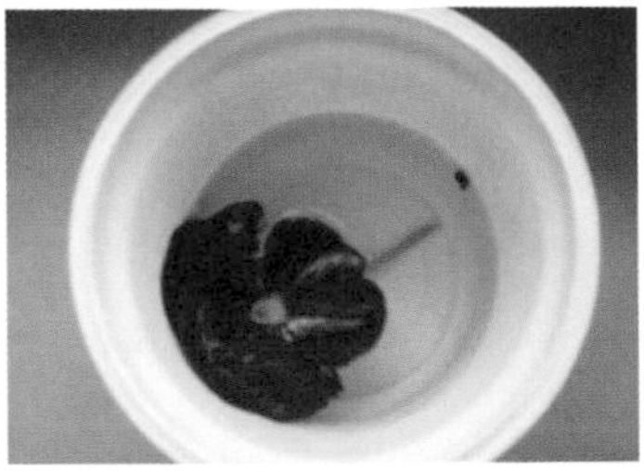

11.15 am 11/04/09: The start of the experiment: the flower dipped in the solution.

7.00 pm 11/04/09: The solution became slightly pink, but the flower has not changed its color.

7.00 pm 15/04/09: The solution became red, the flower changed its color to orange.

Experiment: 1.14
Geranium plant grown hydroponically - aim: to check if geranium can be grown hydroponically and, if so, how long it takes for the plant to wilt after removal of water, whether it can recover and how long this takes.

No changes in the condition of the plant were noted for two days until the absence of water, whereupon reactions began in the lower leaves of the plant.

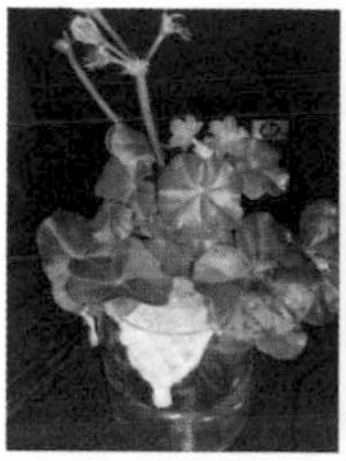

10.15 am 09/04/09: The start of the experiment: roots of the plant immersed in water.
4.40 pm 09/04/09: No change - the plant looks well.

2.00 pm 10/04/09: No change.
10.00 am 11/04/09: Water is removed from the jar.

7.00 pm 11/04/09: Lower leaves have yellowed.

7.00 pm 15/04/09: Lower leaves are yellowish brown, but the plant has not wilted.

Experiment: 1.15
Geranium in soil watered with a solution of blue dye and water - aim: to check if the flower changes color under the influence of tinted water.

11.30 am 09/04/09: The start of the experiment: the plant watered with the tinted water solution.

2.00 pm 10/04/09: No change.

Experiment: 1.16
Geranium in soil watered with a salt solution - aim: to check if the plant wilts when watered with a salt solution (according to an experiment in making a plant wilt, the plant should wilt after 1 hour).

Geranium appears unsuitable for the purpose of this experiment, which should be carried out on a different type of plant whose fleshy leaves store water more easily.

11.30 am 09/04/09: The start of the experiment: the plant watered with the salt solution.

4.40 pm 09/04/09: No change.

2.45 pm 11/04/09: Lower leaves only have wilted, the plant stem is limp, the lower leaves have yellowed. The flower stem appears unchanged.

3.00 pm 11/04/09: Soil has been removed and the plant immersed in water.

2.45 pm 11/04/09: Lower leaves only have wilted, plant stem is limp, lower leaves have yellowed. Flower stem appears unchanged.

Experiment: 1.17
Cut pansy flower in a 30% vinegar solution - aim: to check if the flowers change color under the influence of vinegar (since the experiment with the pansy plant grown in soil was unsuccessful, the question arises whether it is possible to change the color of cut flowers in the same way).

Basil is not ideal for this purpose because its stem is overly delicate and flimsy; potted Basil plants purchased in shops are wrapped in foil which supports the plant foliage but when the foil is removed, the foliage is unable to support itself and droops.

10.30 am 09/04/09: The start of the experiment: roots of the plants dipped in water.
4.40 pm 09/04/09: No change - the plant appears healthy.

2.00 pm 10/04/09: The plant has wilted slightly though this is barely noticeable - the stems are not as rigid as at the start.
10.00 am 11/04/09: Water removed from the jar.

2.00 pm 11/04/09: First signs of wilting (but only leaves wilted slightly, the stems look fine).
7.00 pm 11/04/09: The plant wilted completely.
7.00 pm 11/04/09: The jar refilled with water.

7.00 pm 11/04/09: The plant wilted completely.

2.00 pm 11/04/09: First signs of wilting (but only leaves wilted slightly, stems look fine).
7.00 pm 11/04/09: The plant wilted completely.
7.00 pm 11/04/09: The jar refilled with water.

Experiment: 1.18
Cut Chrysanthemum flower dipped in a solution of blue dye and water - aim: to check if the flower changes color under the influence of tinted water.
Result: Chrysanthemum flower didn't change color

5.15 pm 09/04/09: The start of experiment: the flower dipped in a solution of water and dye.
2.00 pm 10/04/09: No change.

10.00 am 11/04/09: No change.
3.00 pm 11/04/09: No change.

Experiment: 1.19
Chrysanthemum plant in soil watered with solution of blue tinted water - aim: to check if the flowers change color under the influence of water and dye.
Result: Chrysanthemum flower didn't change color

5.15 pm 09/04/09: The start of the experiment: plant is watered with a solution of water and dye (Check in same way if flowers change color under influence of food coloring).
2.00 pm 10/04/09: No change.
10.00 am 11/04/09: No change.

3.00 pm 11/04/09: No change.
10.00 pm 13/04/09: No change.

Experiment: 1.20
Chrysanthemum in dry soil without watering over an extended period of time – aim: to check how long it takes for the plant to recover.

Result: Chrysanthemum can completely recover from dry soil over an extended period of time. The plant was observed over a period of 4 days after first watering and complete recovery was found possible. It has not been tested whether this plant might be grown by means of hydroponics, nor how long it might take to wilt when deprived of water.

10.30 am 13/04/09: The start of the experiment: the plant has been watered.
12.00 pm 13/04/09: Top leaves of the plant visibly stood up, but the whole plant hasn't been restored completely.

12.50 pm 13/04/09: The whole plant has been recovered (all leaves stood up), but the lowest leaves don't look as healthy as before.

3.00 pm 13/04/09: The whole plant recovered completely, it looks 100% healthy.

Experiment: 1.21
Cut Gerbera flower in a 30% vinegar solution – aim: to check if the flower changes color under the influence of vinegar.

Result: Gerbera flower didn't change color under the influence of vinegar solution and it was dead and can not be reversed.

5.20 pm 09/04/09: The start of the experiment: the flower is dipped in vinegar solution.

2.00 pm 10/04/09: The stem of the flower has wilted, its flower is not dead, but its color has not changed. Attempt recovery of the stem's condition.

5.00 pm 10/04/09: The vinegar solution is replaced with water.
10.00 am 11/04/09: The flower did not recover and is dead.

Experiment: 1.22
Gerbera plant in soil watered with a 30% vinegar solution – aim: to check if the flower changes color under the influence of vinegar.

Result: Gerbera flower didn't change color but its leaves turned slightly yellow before completely wilting.

5.20 pm 09/04/09: The start of the experiment: the plant is watered with the vinegar solution.

2.00 pm 10/04/09: No change.
10.00 am 11/04/09: No change.

3.00 pm 11/04/09: The leaves of the plant have yellowed somewhat, but flower color has not changed.

7.00 pm 15/04/09: The plant has wilted.

Experiment: 1.23

Gerbera plant in dry soil which has remained without watering for 4 days and whose flowers have wilted is given water – aim: to check if the flowers recover.

Result: Gerbera flower didn't recover.

12.45 am 11/04/09: The start of the experiment: the plant is watered.
2.00 pm 11/04/09: No change (the flowers did not recover).
6.00 pm 13/04/09: No change (the flowers did not recover).

GENERAL COMMENTS

It was observed that plants grown hydroponically react more rapidly to stimuli (i.e. vinegar) than plants grown in soil:

– lack of water – it is far easier to control wilting of foliage when a plant is grown hydroponically, simply by removing water from its container; obtaining foliage wilting in dry potted soil is time consuming, as potted soil must be deprived of water for some time to obtain a sufficient lack of humidity;

– in adverse conditions i.e. lack of water, vinegar etc., the flowers of most of the plants tested (Pansy, Gerbera and Geranium) die rapidly and are not recoverable or cannot be recovered fully; of all of the varieties of plants test, only Chrysanthemum flowers were found resistant to adverse conditions and all those tested recovered.

– plants with thick, fleshy foliage use their leaves to store water. As a result, they react slower/are more resistant to stimuli (Geranium is extremely resistant to salt and far more resistant to a lack of water than i.e. Pansy);

– all experiments on recovering plants after the application of salt and vinegar were unsuccessful;

– as a network status indicator, the wilting effect of Leopard Lily and Chrysanthemum were successfully tested; the plants either died or lived, but a midway stage was unobtainable.

DECISIONS ABOUT A NETWORK STATUS INDICATOR

After all of these experiments, we concluded that plants used as interfaces are an unsuitable idea for this project, but what else could we use as an indicator? The simplest idea was to use an appliance that plugs into the unit as an indicator itself. Imagine a lamp that kept flashing so much that you could barely use it as a lamp. The longer it could be on means the better network status indicator it is. Eventually we came up with the idea of a 3 step switch, designed so that the user can choose how they want to use the energy. The user will feel the network status only when the unit is in "Selfless" mode where the unit tries to balance the CO_2 sequestered and released into the system.

THE CHOICE OF PLANTS

Relinquishing of the use of plants as indicators considerably widened the range of plants viable for the project. Thus, the short-listed plants must fulfill only the following conditions: be reasonably priced, widely available, suitable for indoor cultivation, easy to grow and of large mass, that is, high carbon sink capacity. Despite its large mass and low maintenance advantages, Leopard Lily appeared to be too expensive and not easy to obtain in the quantities required for the project. Therefore, other options were investigated. One of these was an interest in plants that reduce pollution, not only absorbing high amounts of carbon dioxide, but also other toxic substances.

According to a study conducted by NASA, a number of houseplants clean the air. The list below contains 10 plants that are the most effective in removing Formaldehyde, Benzene, and Carbon Monoxide from the atmosphere:

- Bamboo Palm *Chamaedorea Seifritzii*
- Chinese Evergreen *Aglaonema Modestum*
- English Ivy *Hedera Helix*
- Gerbera Daisy *Gerbera Jamesonii*
- Janet Craig *Dracaena "Janet Craig"*
- Marginata *Dracaena Marginata*
- Mass cane/Corn Plant *Dracaena Massangeana*
- Mother-in-Law's Tongue *Sansevieria Laurentii*
- Pot Mum *Chrysantheium morifolium*
- Peace Lily *Spathiphyllum*
- Warneckii *Dracaena "Warneckii"*[6]

It was decided to use *Spathiphyllum* for "Natural Fuse" since this plant fulfilled all set conditions. It requires humidity, needing to be watered intensively especially in summer and flourishing when its leaves are sprayed with water. It needs a sheltered, semi-shade environment away from direct sunlight. It grows well in loosely packed, nutrient rich soil with good drainage. Optimal temperatures range between 18°C and 21°C (64°F and 70°F).

Experiment: 1.24

Killing system: Spathiphyllum watered with 100ml of vinegar - aim: to check a) if 100ml of vinegar is sufficient to kill the plants, and b) how long it takes for the plants to die.

7.00 pm 12/08/09: The start of the experiment: 1000ml of water is poured into the bottom of the container and 100ml of vinegar poured into the reservoir located at the bottom of the planter.

7.00 pm 13/08/09: The end of the experiment: no change.

Experiment: 1.24.1

Vinegar is poured into the bottom of the container.

Observation continued over a period of 4 days and no change in plant condition was observed.

7.00 pm 13/08/09: The start of the experiment: 50 ml of vinegar poured into the top into container.

After 6 hours.

Experiment: 1.24.2

After 12 hours.

14/08/09: After 18 hours.

Vinegar is poured into the top of the container.

The inference here is that plants absorb water during the day - this test began at 7.00 pm, delaying the start of water absorption. Therefore, if vinegar were added early in the morning, the plant would probably wilt faster.

Conclusion: 100ml vinegar poured into the top (directly into the container) should be sufficient to kill the plant.

2. DEVICES

A. WATERING DEVICES

PWM (Pulse-Width Modulation)
As a network status indicator, we successfully tested the wilting effect of Leopard Lily. The plants either died or lived, but a midway stage was unobtainable. Therefore, it was decided to seek 30% or 80% wilting. In order to regulate the amount of water supplied to the plants, a timed dosing pump controlled by an Arduino was created, which waters the plants at certain intervals. In this experiment water was given to Leopard Lilies in a 2-hour loop, in an attempt to create moderate wilting.

Watering system 1

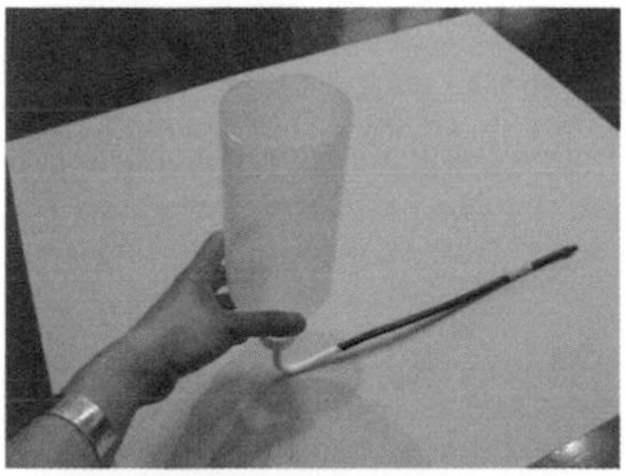

Bottle with cap and sprinkler tube.

Bottle cap with sprinkler tube connector.

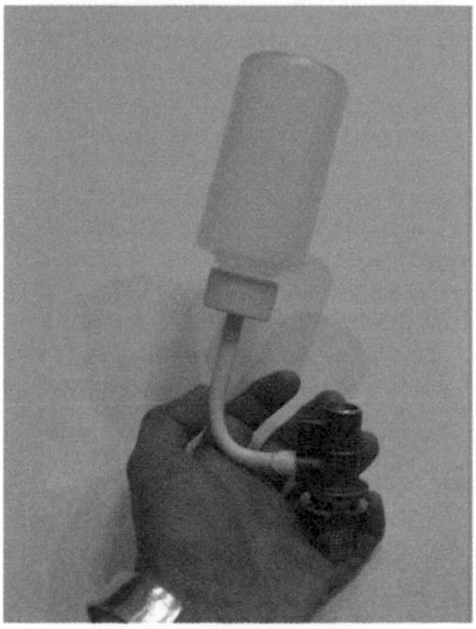

A toilet flush valve attached with solenoid.

Watering system 2
An automatic watering plant pot evidently needs a highly efficient watering system. The "Self-Watering Balconnière" from The Stewart Company is excellent for the purpose of these experiments.

Briefly, water is collected in a tray beneath the plant container, feeding water to the plant via its roots, resulting in less humidity loss and therefore more water retention in the soil.

This planter was chosen for the staging of "Natural Fuse". However, the volume of water given to the plants must be controlled, for which the most energy-efficient method of creating water flow is utilized, namely, gravity.

Watering experiment 01:
An attempt was made to use solenoid to open the bottle cap at the end of the tube, using gravity to move the water. At issue was the need for excessive force to push the cap against the tube to create a water-tight seal. The more power was used, the more power was needed to remove the cap. Further, this system was somewhat complicated to construct.

Watering experiment 02:
This solenoid valve from RS Components was found for £6 and it was hoped that it would perform as required in experiment 01. The results were extremely disappointing, as water dripped rather than flowed, taking about 2 minutes to move 20ml of water. Later, it was discovered that this particular valve needs at least 2.0 bar water pressure to operate, while a low pressure solenoid valve costs £60. Therefore, this option was rejected. Also, the power consumption of this solenoid valve is roughly the same as that used by a water pump at the same voltage, pumping in 5 seconds what the solenoid valve pumps in 2 minutes. Therefore, a solution was sought elsewhere.

Watering experiment 03:
These fuel pumps for RC cars can be purchased for approximately the same price as the solenoid valve to conduct tests. The cylinder version functions much better, is less noisy and consumes less power. Unfortunately, this was the last model in stock in any of their UK branches. Furthermore, there were only 32 of these rectangular models available at stores throughout the UK, while at least 50 were needed for the project. After searching on the Internet, an OEM pump that appears

identical to the RC fuel pumps is located at half the cost. It is produced in China, and the supplier is contacted. Apparently, there is only one component material inside that is different, which should be of no concern, so 55 were ordered and tested on arrival. However, results were extremely disappointing, as the pump could not even move the water. It was discovered that this OEM version was made for a car battery requiring high current consumption, while power supply available was only 500 mA. Given the conclusion of these experiments, the rectangular RC fuel pump from Hobby Stores appeared to be the best option, However, further thought was needed regarding how to move the vinegar.

Watering experiment 04:
A low cost water pump used to administer the vinegar when "Natural Fuse" exhibited in Manchester was tested next. A disadvantage of this pump is that it needs to be submerged. However, the pump functioned well for the vinegar reservoir, which held a minimum of 100 ml of vinegar, sufficient to kill a plant. A 150 ml capacity glass jar of the correct size was found from IKEA, adequate in size for the vinegar as well as the mass of the pump.

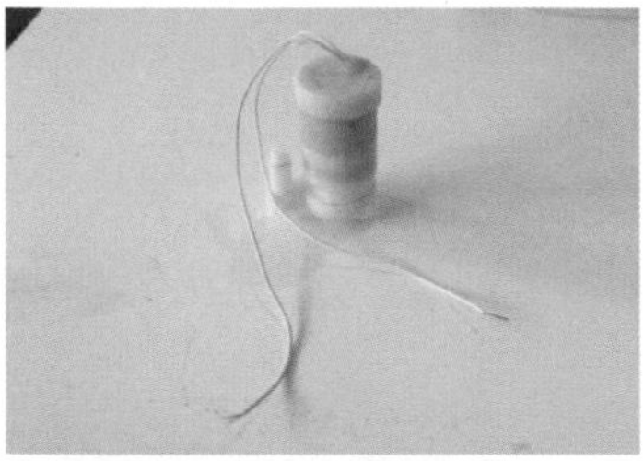

Watering experiment 05:
During one of the tests, a rectangular pump was damaged and it was assumed that it was left to run too long without water. Although the planter and its plant consumed an entire bottle of water in approximately 2 weeks, lack of attention to water levels could be critical.

Therefore, a sensor was attached to the watering tube to test for the presence of water in the bottle. This sensor consisted of a simple wire connected to a resistor and an Arduino board. This functioned well for a day or so, but the end of the probe corroded, giving a continuous reading of the absence of water.

The pump's ability to function without water was also tested. An initial investigation was held, during which the pump was set to run for 10 seconds every hour in the absence of water. The conclusion of this investigation was satisfactory, given that the period of 10 seconds was not extended.

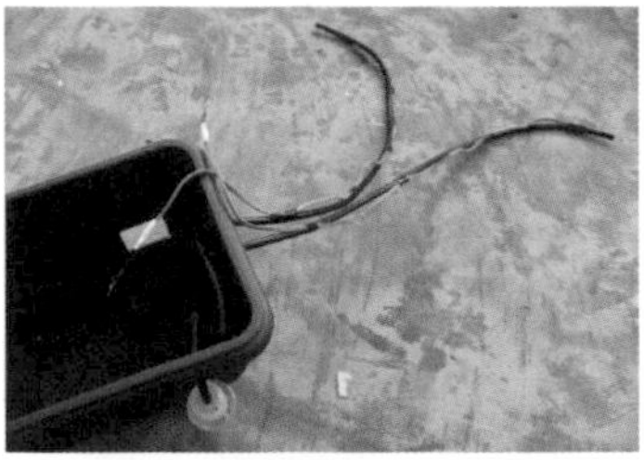

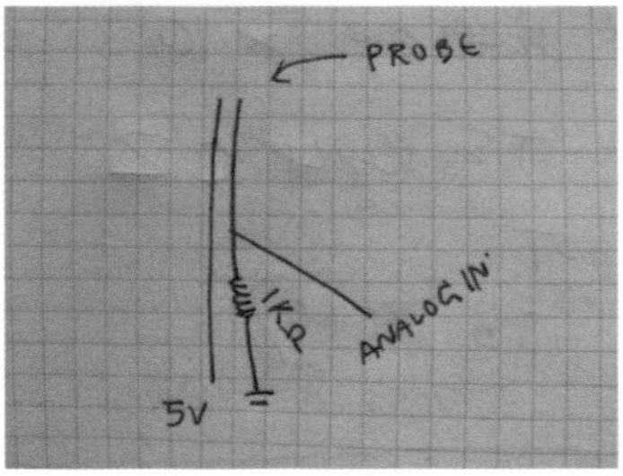

B. ELECTRONICS DEVELOPMENT

Latching Relay
Since energy consumption needed to be minimized, a Latching Relay was chosen. In a normal relay, power is constantly supplied to the relay coil while it is on, which is not at all energy-efficient. The Latching Relay has two coils, which require power only when it is turned on or off, making it an excellent energy-saver, especially when it is turned on for long periods of time."

Issues with the Ethernet shield
An ongoing issue was encountered with Arduino 328 and the official Ethernet shield which functioned only when run with a serial monitor from the Arduino program or by pressing the reset button before running. This problem occurred only with Arduino 328, as the model used did not reset the shield properly at power start-up.

The problem was resolved with special thanks to a user named "agt" from the Arduino forum. The solution was to bend the reset pin of the Ethernet shield so it could be reset separately from the Arduino. Then just one digital pin could be used to reset the shield by pushing it to LOW before running the Ethernet Library.

In this image, pin 9 was used to reset the shield. (Note: disregard the 5V, ground and Vin pins that are also bent).

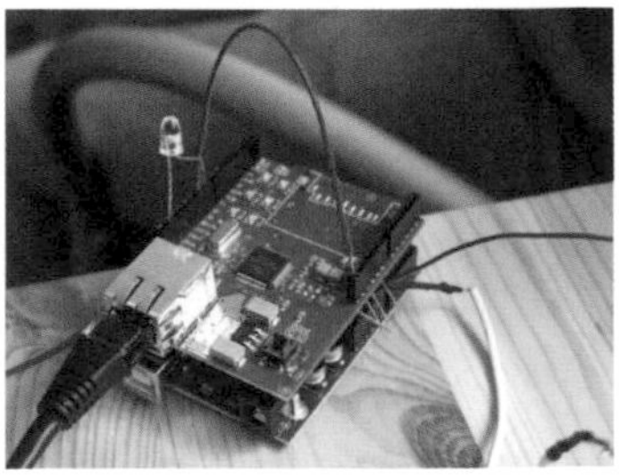

Ethernet shield = power sucker
According to measurements taken, the Ethernet shield consumed the most power in the unit. Arduino consumes roughly 0.12 watts, but the shield consumes 1.8 watts, about 15 times higher than that of Arduino itself. Therefore, it was decided to turn off the shield most of the time, using it only when necessary. The Ethernet shield uses 3.3V from Arduino to operate, so a latching relay between the 3.3V pin from Arduino and the Ethernet shield was added.

From the image, 3.3V pin is connected to the latching relay, then the shield, which functioned adequately for a while. After constructing 25 circuits, 7 Arduinos broke since, apparently, the Ethernet shield drew too much current from Arduino. Assuming that this normally occurred only when both are started up, it was decided to put in place a 3.3V regulator to supply power to the shield (convert 9V external power supply to 3.3V).

From the image, 3.3V regulator in position.

Power consumption measuring circuit
In order to determine the amount of power being drawn by any appliance plugged into "Natural Fuse", a fairly simple circuit was included to determine the approximate current draw. This was accomplished by taking measurements with an op-amp of the voltage drop across a known low-value resistor using the equation Current = Voltage/Resistance.

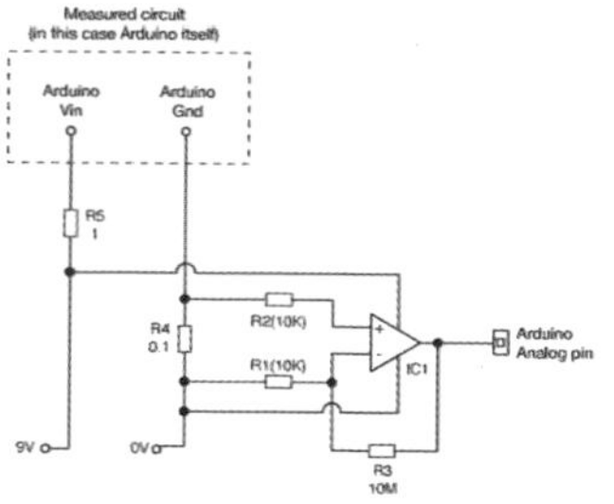

IV. CONCLUSION

The plants used in these experiments may be categorized as follows:

Lack of water
Some plants, like Leopard Lily and Hydrangea react rapidly to a lack of water. However, the main issue is reversibility.

Over-watering
There may be a longer reaction time to over-watering than to a lack of water. Examples of plants in this category are Christmas Cacti and Areca Palm.

Chemical and substances
Multiple tests have shown that some plants react to Ammonia, Acetic Acid, Calcium or Copper. The most common reaction is in a change in color of the flower when reacting to Ammonia.

Fertilizer
Excessive fertilizing can cause minimal leaf burn in Leopard Lily and can also cause yellowing of foliage in Areca Palm.

Color Dyeing of flowers or leaves
Many plants and their flowers respond to food coloring and whether this is reversible is not known as this stage.

Soil PH
Changing the color of Hydrangea flowers by adding Aluminium Sulfate/Sulphur or lime to the soil is popular, although this process may take three months to complete.

Soil salinity/hard water
Leaf color change in many plants such as Gerbera and Freesia is a common result of watering with hard water.

Orientation of the pot to light
A change in direction of the light source may cause Madagascar Jasmine to lose leaves and flowers.

Touch
Mimosa folds its leaves under various stimuli such as touching or moving the plant, unfolding in sunlight and folding at night.

V. INTERVIEW WITH USMAN HAQUE, BY MATTHEW FULLER [10]

Matthew Fuller: The documentation of the experimental stages in the development of the design includes a lot of footage of dead and withering plants as potential plants are tested on how fast they can be killed by the application of vinegar to their various growth media. The idea of sustainable technology tends to suggest a narrative of improvement in which the basic infrastructures of western society can remain untouched, indeed globalized, whilst their modes of production and consumption are to be made kinder and gentler. Your homely landscapes of poisoned soils and houseplant-scaled deforestation at once poses the idea of individual solutions, of ingenuity in handling and testing them, but also perhaps stages it in terms of fundamental and multilayered problems that are incommensurable with contemporary visions of an easy energy future?

Usman Haque: I'm interested in the situation well described by game theory's "prisoner's dilemma". It is sometimes used to explain why it is so difficult for human beings to take coherent decisive action with respect to tackling the issues surrounding the environment and climate change: whoever makes the first move towards tackling global problems in the short term is bound to suffer the most (this is unsurprisingly most often expressed in economic terms). Prisoner dilemma shows how it is quite possible for us to make logical decisions that appear to be in our own interest, but which, when viewed from a global perspective are actually counter to our own interests.

But initiatives like the Grameen bank in Bangladesh and other micro-credit systems have provided intriguing strategies for 'socializing' risk. In the Grameen bank, for example, although individuals take out loans, the community as a whole is responsible for repaying them - this partly relies on peer-pressure with respect to ensuring that individuals repay, but also partly on the idea that there will be a collective attempt to help out an individual in time of need. I'm interested in exporting this kind of approach to the debt that we owe to natural resources.

The point is that there is no 'easy energy future'. We've got to stop trying to sell people the idea that there are obvious ways to deal with the kinds of complex systems that govern both our social and environmental lives. It is often expressed that it is the task of designers to "make things simple for people" - which I find patronizing and counter-productive. If anything, it is the task of designers to show how *complex* things are, and to help build tools for dealing with that complexity (which is the basic function of the perceptual systems we are endowed with anyway!)

Whether it's bio-fuels on the one hand (which for a brief moment seemed to be an 'obvious' solution) or extensive government subsidies (in the UK) for homeowners installing solar panels (which, when you do the math makes little economic sense, and merely makes people feel happy they're "doing" something), we keep discovering that the 'easy' option has detrimental consequences.

M.F.: Systems designed by tacit knowledge and slow custom-based development (such as the evolved designs of unpowered ships and boats) often allow, within a general approach of precautionary over-engineering, for certain components, usually the cheapest and easiest to replace, to be the first that will break under particular stresses, thus saving the larger structure. Within a sailing ship, these might be smaller cords attached to the larger stays holding a sail in place. These would snap if a wind suddenly became too strong, in a way that might otherwise damage the mast or sail. Power would be lost, but the core parts of the structure would remain undamaged. This 'design to fail' approach is quite different from the imperative for 'graceful degradation' often found in computing and HCI, where crashes are seen as abhorrent and problems are sublimated. But it is also different from the 'fail-free' design approach, such as those developed on the bases of highly engineered but mathematically driven and ostensibly optimised design which imagines problems can be simulated out. In consumer electronics the fuse, embedded in the plug, is of course the part designed to fail if an electrical surge is encountered. I wonder with this project if there is a more general ethic of brokenness that you subscribe to in design?

U.H.: This is an intriguing way to look at it, and I hadn't really considered "Natural Fuse" in those terms. But, certainly, embedded in the core concept of the project is the idea of the 'canary in a coal mine' - using proxies for ourselves that break earlier and less expensively (in economic, social and ecological terms) in order to make it clear before greater damage is done.

There is also an aspect of the project, not, I should say, carefully considered, that concerns the use of plants: if we had killed an animal instead of a plant, that would be a lot more uncomfortable for people (and they probably wouldn't have wanted to take on the responsibility, considering the life of the animal is in the hands of someone else).

So using plants (apart from their carbon-capturing aspect) means that we can conveniently offer people something that they won't need to "worry" about too much, but which, nonetheless, grows on them - when you adopt a plant, over time you become attached to it. So, surreptitiously, perhaps, we've got something into your home that you didn't think you would care too much about, but which you *do* actually begin to worry about and have concern for its well-being. Plants seem non-threatening (in the sense of responsibility), but ultimately become quite important to people. And this is intriguing since we tend to think that you can kill plants indiscriminately (in a way that is not morally acceptable for animals), even though they may be extremely good at helping us survive. A colleague has referred to this as "horti-torture"!

M.F.: One thing that is notable in the project is that it reverses the genetic engineering scenario of the plant being switched on and off at the chromosomal level by technologies working on biological material through the metaphor of information. In this case, electronic systems are shut down by organic material. Does "Natural Fuse" suggest some convergence of the informational view of life and a more organismic or ecological sense?

U.H.: Partly, yes: but only because at heart I'm interested in systems, and more specifically I'm interested in 'coupling' systems. Most of my work looks at how we can couple human and non-human (I don't say "natural" because that implies that humans are not "natural") systems; electronic and social; ecological and economic.

That's also why, when we introduce "Natural Fuse" in a city, we try to encourage an economy of plants. I want to disrupt the conventional economic approach, where money is used because it's convenient. In "Natural Fuse" people rent the units by paying with plants that they bring to the exhibition or the store - they actually have to bring 5 or 10Kg of plant material which they leave behind. This is enough of an investment in time and effort that they must really want to participate.

In New York, they were also able to rent in US dollars, ultimately donated to the Bronx River Art Center, but the rental fee was high enough to act as a disincentive and make paying with plants much more attractive.

Upon returning the "Natural Fuse" unit, they actually get their plants back (so in conventional economic terms they rented it for 'free'): in fact the way they have "paid" is by lending us the carbon capturing capacity of the plant they left behind (which is applied to a very slowly brewing cup of coffee: it takes many dozens of plants growing for a long time to offset the carbon footprint of making a single cup of coffee!)

M.F.: An underlying argument of the project is that design produces social architectures. Every object stages a set of more or less stable relations between infrastructures, resources, ecological processes, organisms and technicities that imply or require forms of community, of participation, of intelligence, that in a certain way articulate the idea of the perfect user/s, or provoke encounters with the abstractions, ideas and actual forces that are perhaps sometimes occluded in certain kinds of design. To say this another way, "Natural Fuse" brings assumed ease of function, for the human user, painfully to the fore. Is this a kind of design for inhibition or for knowledge?

U.H.: I'm not really trying to "communicate" something with "Natural Fuse" - it's not that I want to say, "you must conserve energy because otherwise we will all suffer". I think such strictures are counter-productive: we just don't like being told we must do something. So it's not about communicating.

Much more than energy issues, the project is primarily an experiment in the structures of participation: how can one design a system in which available options are increased (e.g. you don't just have "off/on", but you have "off/selfless/selfish"), while making it possible (and more likely) that people will make decisions that benefit the community as a whole. (See reference to prisoner's dilemma above). I can't say that it's necessarily been a total success: we usually leave the actual "kill" function switched off for the first few days after launch, simply because it takes people a while to fully grasp that they may be killing other people's plants on the basis of their own decisions. And, interestingly, the demo unit left in the store or exhibition, which people feel no "ownership" of - was constantly left by visitors in "selfish" mode, to the extent that we had to remove it from the network calculation because otherwise it would have always anonymously killed other people's plants.

Clearly, as a designer, I have some idea of what I consider desirable goals for the kinds of things that I hope people do. I would like people to act in a way that benefits the community as a whole. And it finds me unusually optimistic: I feel that a project like "Natural Fuse" shows that people *can* make altruistic decisions in order not to harm people they don't know.

In terms of participation, the point is to involve people actively in the processes of decision-making and *also* in the processes of carbon-capturing/energy reduction.

One of the major problems that I see in the so-called climate "debate" is that we are constantly told that there is plenty of "data" out there for us to consume and process, and that conclusions should be obvious or self-evident. But it is very difficult for ordinary people to form their own opinions about environmental and energy issues - they are confronted with so many dozens of valid explanations, visualizations and extrapolations of the data from a variety of authority figures (politicians, scientists, media figures), but much of it is conflicting or contradictory. Authority figures try to tell them what to believe - but the authority figures don't all agree, which means people just opt out.

I think it is vital for people to be able to participate in the process of evidence gathering: partly so that they can question the 'standards of evidence', partly so that they can become part of a solution, but also so that they can understand the methodological limitations to any data-acquisition (and carbon-capturing) process. In Natural Fuse when a plant dies any carbon sequestered during the growth period is, in the absence of continued sequestration (e.g. by sealing it deep within the earth), soon released back into the atmosphere. A zero-sum situation depends entirely on where the arbitrary boundaries of the system are drawn. So what might you do with your plant? Eat it? Bury it? Weave it? Of course eating it results in carbon dioxide output from the body (exhalation, excretion, etc.); burying it takes a lot of energy; weaving it might be an option - but that becomes very "object" or "product" oriented.

It is important to understand the cascading consequences that sets of decisions can have: first at a local level and later at a global level.

NOTES

1. Dr. B. C. Wolverton How to Grow Fresh Air, 50 houseplants that purify your home or office (Penguin Press, 1997)
2. Ibid.
3. "Parts per million"
4. http://www.akcjodynamika.alte.pl/teksty/teksty.php?tekst_id=006)
5. http://science.howstuffworks.com/carbon-capture.htm/printable
6. http://www.zone10.com/nasa-study-house-plants-clean-air.html
http://www.scribd.com/doc/1837156/NASA-Indoor-Plants
7. Arduino is an open-source electronics prototyping platform based on flexible, easy-to-use hardware and software. It is intended for artists, designers, hobbyists, and anyone interested in creating interactive objects or environments.
8. Arduino 328 is The current basic board, the Duemilanove, uses the Atmel ATmega328 microcontroller.
9. An official Ethernet shield is a board that can be mounted on top of the Arduino board that allows the Arduino board to connect to the internet. There are a number of versions of this shield released by other companies as well.
10. Networking Overload, with Potplants

We would like to thank Gregory Wessner and Mark Shepard for making this project possible, and for amazing support and help from Thumb's Luke Bulman and Jessica Young during exhibition preparation. Huge thanks also go to Austin Houldsworth and Kanittha Mairaing ('Goong') for their enormous contribution to the work. Finally, thanks are due to Natalie Jeremijenko for invaluable advice on plant-based carbon capture and local material procurement.

REFERENCES

— http://www.easybloom.com/learn/overview.html
— http://gizmodo.com/gadgets/gadgets/selfwatering-iv-plant-pot-185538.php
— http://www.cheapvegetablegardener.com/2009/01/fully-automated-computerized-grow-box.html
— http://www.botanicalls.com/
— http://mhi-inc.com/Converter/watt_calculator.htm
— http://www.instructables.com/id/Garduino_Gardening_Arduino/
— http://www.instructables.com/id/Make_an_automatic_plant_light/
— http://www.greenergadgets.com/index.php/design-competition/
— http://www.core77.com/greenergadgets/ientry.php?projectid=22
— http://www.greenergadgets.com/index.php/design-competition/
— www.edukacja.warszawa.pl/plik.php?id=2105
— http://dp.idd.tamabi.ac.jp/bioart/
— http://www.woollypocket.com/intro.php
— http://www.re-nest.com/re-nest/green-tours/green-tour-matthew-and-emmas-eco-environment-084775
— http://www.gardeners.com/Living-Wall-Indoor/37-085RS,default,pd.html
— http://www.earthbox.com/
— http://www.autopot.co.uk/
— http://www.label.pl/po/pomiar_co2.html
— http://www.wentylacja.com.pl/ciekawostki/ciekawostki.asp?ID=4145
— http://aeris.eko.org.pl/?dz=5&poddz=3&str=2&lang=pl
— http://kopalniawiedzy.pl/dwutlenek-wegla-atmosfera-odpornosc-rosliny-szkodniki-Evan-DeLucia-4553.html
— http://www.akcjodynamika.alte.pl/teksty/teksty.php?tekst_id=006
— http://www.nasa.gov/vision/earth/environment/aerosol_carbon.html
— http://www.eurekalert.org/pub_releases/2006-04/uom-hyb041206.php
— http://www.erasecarbonfootprint.com/treeoffset.html

— http://www.eltlivingwalls.com/buynow1.php
— http://ezinearticles.com/?Air-Movement-to-Gain-Control-Over-Temperature,-Humidity-and-CO2&id=685861
— http://edition.cnn.com/2009/TECH/science/04/22/plants.pollution/index.html
— http://www.erasecarbonfootprint.com/treeoffset.html
— http://www.zone10.com/nasa-study-house-plants-clean-air.html
— http://www.scribd.com/doc/1837156/NASA-Indoor-Plants
— http://science.howstuffworks.com/carbon-capture.htm/printable
— http://dsc.discovery.com/news/2008/12/04/carbon-sequester-tech.html
— http://dsc.discovery.com/news/2008/12/04/carbon-sequester-tech-02.html
— http://www.gaiainstituteny.org/Gaia/GaiaSoil.html
— http://convertedorganics.com/index.php/Product/
— http://www.swiatkwiatow.pl/skrzydlokwiat---spathiphyllum-id487.html
— "Uprawa roslin ozdobnych." Pod red. H. Chmiela, Panstwowe Wydawnictwo Rolnicze i Lesne, 2000. (Cultivation of Ornamental Plants, Edited by H. Chmiel, National Agriculture and Forestry Publishing House, 2000)
— How to Grow Fresh Air, 50 houseplants that purify your home or office, Dr. B.C. Wolverton, Penguin Books, 1997

URBAN DIGESTIVE SYSTEMS

TRASH TRACK

MIT SENSEABLE CITY LAB:
DIETMAR OFFENHUBER, DAVID LEE,
MALIMA WOLF, LEWIS GIROD,
AVID BOUSTANI, JENNIFER DUNHAM,
KRISTIAN KLOECKL, EUGENIO MORELLO,
REX BRITTER, ASSAF BIDERMAN,
CARLO RATTI

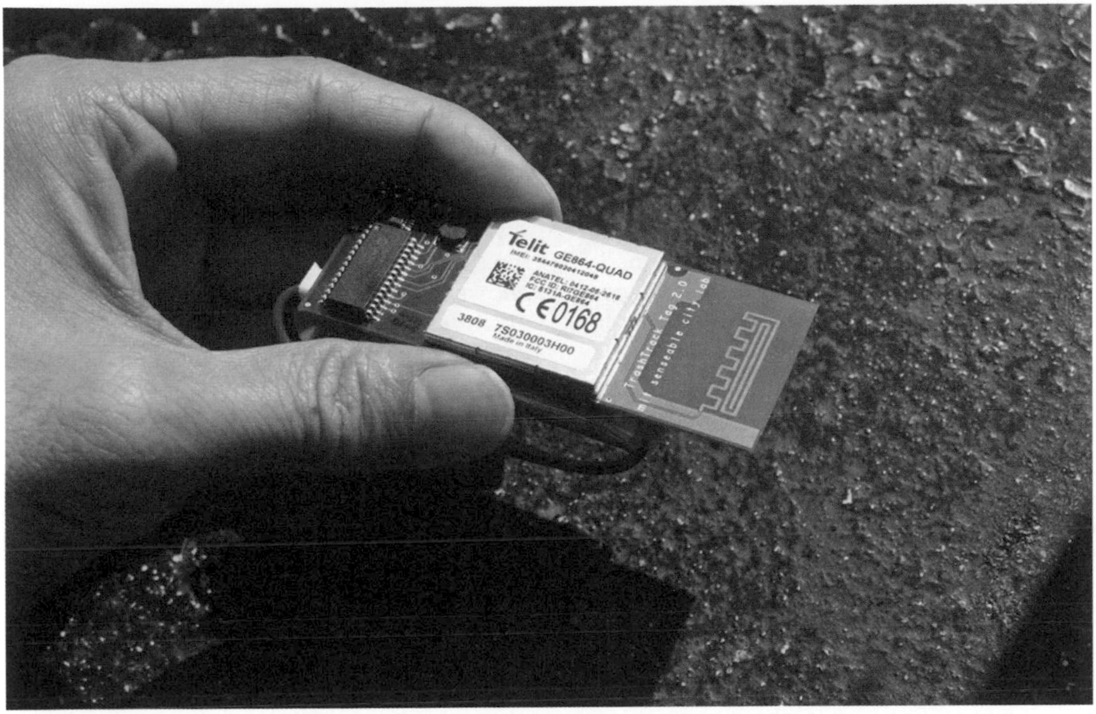

An active location sensor for tracking waster items, developed at MIT.

Introduction

On a warm summer morning in August 2009, Flora steps off a bus in front of the Seattle Central Library, with a ratty old sneaker in hand. Weaving her way through a motley crowd of college students, families, and professionals gathered at the library entrance, she notices that everyone is carrying a unique object; one person has brought a laptop, another drags an empty propane tank, a young girl clutches a stuffed bear.

Eventually, Flora finds a group of MIT researchers huddled over a plastic tarp, and asks them, "Is this where I can track my trash?"

***

Trash Track began with a simple idea, to understand where our garbage goes once it has left our sight. Technology now enables us to track the movement of any man-made object from its creation to its disposal, and the systems governing where our trash ends up are both incredibly complex and rarely understood. Waste travels an elaborate network of transfer stations, landfills, and reprocessing plants, by truck, train, boat, and plane, and is constantly rerouted by regulations and markets. We wanted to understand how well this infrastructure functions, and how individual actions like throwing away a mobile phone translate into wide-scale consequences.

***

The researchers pepper her with questions about herself and trash collection in her neighborhood, and are curious about why she brought a sneaker. Flora talks about its sentimental value, having worn the shoe when running her first marathon. A nearby volunteer jokes, "It would be fun to see how many more miles this sneaker will be able to travel on its own!"

One of the researchers takes a photo of the shoe, inserts a small electronic device into the toe, and pours in some quick-setting foam until the device is concealed from view. After a minute the foam hardens completely, and she hands the shoe back to Flora. "Please throw this away at least ten blocks from the library," they instruct her.

Flora returns home, stopping to toss the shoe into a dumpster outside her apartment. Once inside, she sends the MIT researchers a quick email detailing where she threw away the shoe and at what time. Her work is finished. Trash Track has begun

***

On paper, the experiment design was straightforward: create small electronic 'tags' that could automatically communicate their position back to us, embed the tags in pieces of trash, and disperse them in the city, observing their movement over time. In practice, Trash Track turned out to be quite complicated, and its results raw and open to interpretation. From this data we sought to achieve two goals: analyze how effective waste removal is today, and show the general public how these systems work through maps and visualizations.

***

In late September, Flora spots an update on her Facebook page, announcing that the Trash Track exhibit is now live and publicly viewable at the Seattle Central Library. That weekend, she makes her way back to the library. At the installation, she sees two projectors, cycling through pictures of trash and maps of Seattle, with animated dots and lines darting across the cityscape. For a while, she is transfixed by the video on one screen, showing the inside of a recycling plant and how workers manually sort mountains of trash on conveyor belts.

Turning her attention to the other screen, she recognizes a picture of her shoe hovering over her apartment. A bright line extends from the building, tracing the path of the shoe as it traveled across Seattle, south to Portland, Oregon, and veering east to rest somewhere along the Washington-Oregon border. Small labels pop up on the screen, indicating places where the shoe might have stopped along this path: a Seattle-based recycling/transfer center, a train station strewn with shipping containers, industrial facilities near Portland International Airport, and the Columbia Ridge landfill in rural Oregon. She is surprised by how far the shoe has traveled since she threw it in the dumpster.

***

By having our trash "talk" to us, we sought to connect people with an urban infrastructure that is both ubiquitous and inscrutable. We had to demonstrate how this information could inform infrastructure planning at the city, regional, or international scale, as well as trash disposal decisions at the individual, human scale. In doing so, our research raised questions about the logic of waste removal, the environmental value of information, urban sustainability, and visions for future cities. As a result, Trash Track has sparked debates both in public and private venues that are just as important as its engineering accomplishments.

***

A young man standing nearby is also observing the exhibit, and asks Flora what it is about. Flora explains how she got involved with the Trash Track project, donating her shoe to the cause and helping deploy the tag into the waste stream. She even remarks that, at over 300 miles, her shoe has traveled at least ten more marathons since leaving her home.

They talk about some of the other traces projected on the screen, plastic bottles and computer monitors and paint cans, and speculate as to why some items seemed to travel so far. The young man seems more interested in e-waste, as he works for a company that sells cell phones and batteries, and he notes that cell phones are some of the farthest traveling items. He wonders aloud if there might be a more efficient way to recycle these devices locally.

Flora disagrees, replying, "I'm not sure if I want a factory dealing with those kinds of chemicals so close to where I live..."

Vision

Drawing the connection between my garbage and where it goes creates a sense of responsibility. Hopefully it will help me bridge the gap between my consumer choices and making the planet a little cleaner.

— Trash Track volunteer survey

***

A popular urban myth claims that the Great Wall of China is the only man-made structure visible from space. In truth, this would ignore an even larger man-made structure, the 2,200-acre Fresh Kills Landfill on Staten Island, once the principal garbage dump of New York City for over half a century. Today, Fresh Kills is a Superfund site, shut down to new incoming waste, contaminated with hazardous substances, and continuously monitored by the Environmental Protection Agency at great expense. New York's garbage must now go farther away to be buried.

Waste removal remains a hidden infrastructure of our cities; we consider it functioning well when it interferes as little as possible with our everyday activities. Yet, waste generation continues to rise each year, and the fates of the unwanted materials moved by this system are now understood to have serious impacts on our health and safety. When landfills reach capacity and fuel costs rise, the practice of sending our trash away from the city for disposal becomes increasingly unsustainable. As environmental concerns gain traction, citizens are raising questions about the workings of the waste removal system, especially regarding the efficiency and ecological benefit of recycling programs. Does it really help to sort, transport, and reprocess our trash? Are there fundamental flaws in our regulations, in our infrastructure?

***

Facilitated by new technologies, the Sentient City talks back to its citizens, and will answer these questions and more for us. Aspects of this scenario are already starting to materialize: with the proliferation of mobile, location-aware technology in the form of cell-phones, navigators and other personal devices, an unprecedented amount of information is constantly generated. Creating new services for these emerging platforms, cities and public institutions have begun to share their datasets and provide public gateways for accessing this ubiquitous information in real time. By law, public information had always been accessible to everyone, but it is important to distinguish between haphazard access to raw data and a more active, streamlined distribution. Real time information presented in machine-readable format can be reprocessed, recombined, and reflected upon, finding new purposes limited only by the imagination of the community.

The Bay Area Rapid Transit organization, for example, provides access to real time vehicle positions, schedules, and alerts about its service, making San Francisco one of the first cities to boast a subway with an Application Programming Interface (API) (Bay Area Rapid Transit 2008). Using this data source, individual developers, the industry and the public sector have built an infrastructure of mobile applications, web mash-ups and public displays that connect the information from the subway system to many aspects of daily life.

But the Sentient City does not only "trickle down" information provided by institutions and public services. It also offers citizens new ways of getting involved and providing input. Increasingly, cities make use of this approach in order to gather information directly from their citizens. New York's "311 online" project allows citizens to complain about rude taxi drivers, suggest improvements to public services and collect information about the condition of urban infrastructure (New York City Government 2010). All this can be accomplished directly using smart phones, instantaneously, on location. The fact that almost all of the data generated by these mobile services comes with location metadata is often called 'Hyperlocality,' which represents the organization of digital information strictly by the geographic coordinates where it was generated. Hyperlocality is thus crucial to interpreting and contextualizing information generated in different places.

In some situations, the Sentient City can even bypass public institutions completely. Situated technologies create information ecosystems constituted entirely by the interactions between citizens. These ecosystems do not rely on central authority to mediate information exchange. Instead, mutual reviews and assessments by the participating individuals can establish a distributed system of reputation and trust. Platforms such as Ushahidi combine mapping, real-time data feeds, news aggregation and other tools for collaboration (Okolloh 2009); originally designed to provide a quick and self-organized way to coordinate actions in a catastrophic situation, these kinds of platforms also have great potential for planning and improving everyday services and infrastructures without the weight of bureaucracy.

***

As the proportion of world population living in cities continues to rise, so too does the importance of well-functioning urban infrastructures, services, and ecologies. Monitoring and maintaining the health of these systems is crucial, because the stakes have never been higher. The complexity of controlling congestion, crime, pollution, and other urban ills increases with population; an infrastructure failure can produce catastrophic results. One way to manage this complexity on the scale of the city is through ubiquitous computing and sensing, made accessible to the public.

The technologies that enable this go by many names: 'Blogjects,' 'Smart Dust,' the Internet of Things. These concepts are founded upon Mark Weiser's vision of "The Computer for the 21st Century," a ubiquitous logic deeply embedded in everyday environments and actions (Weiser 1991). His vision liberates information from the virtual space behind the screen. Instead, technology is seamlessly embedded in every object, effortlessly integrated into daily routines. Arranging smart objects on a table triggers meaningful computational interactions, just as command sequences do on a traditional computer. Using sensor data, phones automatically recognize social context, such as a meeting or film screening, and adjust their behavior accordingly.

This idea extends to scenarios where every object can sense relevant information about its environment, infer the situation, and spontaneously communicate with nearby devices or broadcast its state to the world. Julian Bleecker dubbed these blogging objects, or 'blogjects' (Bleecker 2005). Alternatively, as sensing devices miniaturize and distribute pervasively, they become smart dust, an invisible, intelligent infrastructure for real-time environmental sensing. Through such transformative technologies, the Sentient City can meet our growing demand for rich, accurate, timely information.

Smart dust, in particular, was a useful starting point for investigating the movement of garbage. We scattered our location sensors into the removal chain like dust particles in the wind, with no hope of retrieving them afterwards. The sensors journeyed within heaps of trash, detected their own locations, reported back to our servers, and allowed us to observe the unseen infrastructure they inhabited. Eventually, their batteries depleted and their circuits were damaged from wear and tear, they stopped speaking and became indistinguishable from the surrounding trash.

***

In 2008, journalists from Greenpeace investigated rumors of electronic waste exporting, a legal practice in the United States, but outlawed in most countries of the world as a result of the International Treaty of the Basel Convention. The journalists embedded GPS sensors into television sets that were broken beyond repair, and brought them to a local recycling facility in the UK. Despite legislation banning the movement of e-waste between nations, the defunct televisions were tracked all the way to Nigeria, where they were likely illegally dumped.

Waste removal is a complex process, subject to myriad regulations and multiple exchanges that are difficult to track. Potential for fraud has always existed in this tangled web, driving a long history of organized crime control over the industry in various parts of the world. Most often this disproportionately affects poorer areas over wealthier ones, raising questions of environmental justice when the consequences threaten public health. Globalization has made these transboundary issues, as people attempt to cheaply export hazardous waste to countries with lax regulations, instead of complying with local standards. Paradoxically, the very regulations that should prevent misuse have generated a grey market for waste, where profit can be made by bypassing legal procedures.

The sheer volume of garbage that we generate also threatens sustainability. Goods consumption and garbage production are constantly rising, yet safe and affordable landfill space is becoming harder to find. In addition, most small regional landfills at the urban fringe have been shut down, as cities have expanded and pushed out these practices. As a result, waste travels much further than before, using more energy and producing more emissions. Furthermore, closed landfill sites often carry toxic legacies that require remediation and constrain later land use; thus hazardous waste disposal becomes especially costly and vulnerable to fraudulent practices.

***

To ensure responsible waste removal, the proximity principle, as outlined in the Basel Convention, demands that waste is disposed of as close as possible to its source (Kummer, 1999). However, waste removal chains are often inadequately monitored to ensure enforcement. Such chains can involve many different companies, but very little information is passed between them; a standardized information model for tracking waste does not exist. As a result, current procedures for monitoring the process rely on paper protocols and voluntary evaluations of facilities, a system built almost entirely on trust.

As the "following the e-waste trail" project by Greenpeace shows, technology can overcome many of these obstacles (Greenpeace International 2008). Active GPS-based location sensors can effectively monitor individual waste items anywhere in the world, highlighting potential abuse. Because sensors are irretrievable once thrown away, GSM (Global Standard for Mobile Communications) and other mobile phone networks are important infrastructures for real-time communication. Online databases collect and store this data remotely; advanced analytical and mapping software allow us to make sense of massive amounts of location data and pick out patterns from the noise.

There are also low-tech methods of tracking the motion of refuse, like relying on many volunteers to act as temporary sensors. A famous naval spill of 29,000 rubber toys in 1992 enabled a long-term study of ocean currents in the Pacific. Oceanographers Ebbesmeyer and Ingraham recruited local residents, visitors, and beach workers to recover toys being washed ashore in Alaska. Based on the locations of around 400 found toys, scientists refined their models of oceanic currents and correctly predicted the trajectories of the remaining toys (Ebbesmeyer et al. 2007). This demonstrates what can be achieved by engaging a community even without expensive tracking technology.

Beyond the tracking process itself, there are important precedents for communicating our observations to the public. One relevant prior work is Eric Paulos' Jetsam project (Paulos and Jenkins 2005), which sought to improve public awareness of waste issues. They deployed a trash bin in public space that was equipped with various sensors and network connectivity. This 'augmented trash bin' gave feedback to its users by displaying statistical information about collected waste via projected visualizations. Such real-time feedback can deliver information about waste disposal in an intelligible, resonant way, and could potentially drive greater awareness and behavioral change.

These examples are at the core of what we are after. Yet, Trash Track goes beyond all of the above approaches, using pervasive technologies and volunteer mobilization to expose challenges of waste management and sustainability to the world. It builds on previous work by the SENSEable City Lab that explores how the increasing deployment of sensors and mobile technologies radically transforms our understanding and description of cities. The project is an initial investigation into understanding the removal chain in urban areas, and represents a model for change that is taking hold in cities: a bottom-up approach to managing resources and promoting environmental awareness. Trash Track hints at an important, but little discussed aspect of the vision described as the Internet of Things. As argued by the novelist Bruce Sterling, we can direct every object to the optimal reuse scenario and ultimately achieve a condition with zero waste, if we know where all things are in the world (Sterling 2005).

Technology Design

The whole project has made me much more optimistic about our ability to find solutions to big, complex problems. Seeing how the people involved in the project are using their creativity and scientific expertise to tackle something as mundane yet complicated as trash disposal is very encouraging.

— Trash Track volunteer survey

***

The Trash Track project was conceived in the summer of 2008 as a proposal for the Toward the Sentient City exhibition, organized by the Architectural League of New York. Rex Britter, a visiting scientist with the SENSEable City Lab, suggested the idea of tracking garbage in the city in order to better understand the collection of waste — which Bill Mitchell dubbed the 'removal chain' as the counterpart to the supply chain — and eventually improve its logistics. Tracking trash intrigued us both as a window into its environmental impact on the city and as an extension to the Lab's past work in diffuse pervasive digital sensing to explore the physical context of the city.

The Trash Track concept also raised questions and concerns in its developmental phase. Was the inevitable deposit of potentially toxic tracking electronics into landfills contrary to the environmental goal of the project? Any demonstration of this technology would inevitably produce a small amount of electronic waste, but the knowledge and information obtained by tracking waste products could provide enormous benefits outweighing the risks. Thinking ahead to the future, it will be possible to carry out projects like Trash Track in more environmentally friendly ways. Rapid miniaturization, organically based batteries, and self-powering methods will all contribute to the greening of electronic sensing devices.

Evaluating different materials for protecting the sensors from the physical impacts inside the waste stream.

***

While our initial concept of tracking trash was relatively simple, a number of constraints complicated development and implementation. Due to the wide variety of waste items, materials, and treatment processes, we needed a large number of tags to track a representative sample; distributing ten or twenty tags would yield only anecdotal results. It was also impractical to try to recover and reuse tags once they entered the waste stream. For both of these reasons, it was crucial that the tags were obtainable at very low cost.

The range and variety of trash paths made them impossible to predict a priori, so it was critical that the tags were able to self-locate and transmit their location information from anywhere in the country. Finally, due to the potentially long travel time of trash, our sensors required long-lasting battery performance that was not easily met by off-the-shelf commercial solutions. Most available devices have to be recharged every few days; when potentially following garbage for months, this was not feasible.

Given the costs of active sensors, passive RFID (radio-frequency identification) tags seemed like an attractive alternative. RFID was already commonly used to track items in retail supply chains; the low price of these tags made them cost-effective to deploy in large numbers. However, as a near-field localization technology, RFID would have required an additional network of tag reading devices, deployed at every potential stop and destination in the waste stream. Besides the prohibitive cost of building this network from scratch, an RFID solution would also be unable to track the most interesting cases of trash, those that went astray from expected paths.

Therefore, active-reporting infrastructure was critical to tracking object movement in an unconstrained system like the waste removal chain. We chose a tracking technology based on the GSM cell phone network, since it not only provided a solution for location sensing, but also offered a communications channel for reporting the sensed locations back to us. By identifying cell towers in the proximity to the device and triangulating their known coordinates, GSM could provide coarse-grained location sensing. In addition, the GSM network allowed sensors to actively communicate back through those cell towers.

***

Two generations of Trash Tags were developed at MIT by utilizing an off-the-shelf GSM data modem chipset, microcontroller, motion sensor, and a custom printed circuit board

Different waste items waiting to be tagged, contributed by Eatonville High School students.

with an integrated trace antenna. After gaining deployment experience with the first generation tags, a new design was developed prior to the second production run. This development process focused on simplifying tag design in order to (1) reduce power requirements to reduce battery size, (2) shrink form size using a low-cost, small antenna design, (3) consume fewer electronics and improve environmental performance, and (4) minimize packaging without sacrificing the robustness needed to survive impacts in the removal chain.

To minimize power consumption, the tags were equipped with an algorithm that would awaken the tag upon detection of motion by the motion sensor. Tags would scan all channels and bands for cell towers; locate the strongest twelve cell sites nearby and store their identification information. Subsequently, the tags compressed tower reports into Short Message Service (SMS) messages and periodically sent them to a server; low frequency of communication was critical to maximizing battery life. In terms of environmental performance, the second generation of tags was lead-free and used components containing very little or negligible hazardous substances. As a result, the GSM tags complied with environmental standards such as the European Union's Restriction of Hazardous Substances Directive, designed to reduce the disposal of hazardous substances in electronic equipment.

The final deployment used tracking devices developed separately by Qualcomm. These took advantage of GPRS and GPS technology to deliver more accurate location reports. The Qualcomm tags were similar in size to previous generations, and had a sleep mode for conserving energy, a crucial feature for our experiment. Rather than activating on motion, the sensors were configured to report their location every three to six hours; we chose several different reporting cycles to balance between the need for longer battery life and for more detailed traces in space and time.

Evaluating tag performance in the field revealed three important limitations: battery life, failure rates and accuracy of the location reports:

— It was difficult to clearly define expected battery lifetime because of the varied nature of the conditions experienced by individual tags. Since the sleep mode is very low power relative to when the cellular module is active, the lifetime was strongly affected by the amount of time that the tag was actually in motion, and by the algorithm for turning on the system in response to vibration. Based on our data, we estimated that our tags had sufficient energy to operate for 20-30 hours when constantly in motion, and from 3-6 months in sleep mode.

— We noticed that about 20% of the tags either reported very short traces or failed to send a report at all. Tags could have failed for any number of reasons: destruction during the waste removal process, arriving somewhere the transmission signal was blocked, hardware malfunctions of the tracking device, or human errors resulting from volunteers not disposing of the tagged item. We found that failure rates depended heavily on the packaging strategy and type of trash; battery failure turned out to be less of a problem than physical destruction.

— The location reports we received had limited resolution in time, due to the power-conserving measures explained earlier. However, the resolution was sufficiently high for inferences about the route and the mode of transportation. The accuracy for GSM based localization can be estimated at about 250 m. GPS localization significantly improved this accuracy. However, since the GPS signal required a clear view of the sky, this feature was not always available.

Deployments

The blend of technology, information systems, and my own lack of concrete knowledge about where trash goes were the primary catalysts. A strong feeling that I wanted to influence others in the Seattle area to get involved at the local level was a secondary — but still important — motivation.

— Trash Track volunteer survey

***

Over several phases of the Trash Track project, we distributed a total of 3,000 sensors in the cities of Seattle, New York, and London.

The initial field test of trash tags took place in June 2009 in Seattle, Washington. Our primary goal was to test the tags' performance once they had entered the waste removal stream. We also wanted to sample data from a wide distribution of disposal locations across the city. Driving to these target locations, we found many objects simply lying around on streets and empty lots. Examples included beverage containers and newspapers, toys and batteries, cell phones and computers, appliances large and small, furniture and clothing, and hazardous materials and car parts.

We tagged each of these found items, trying out different techniques for attaching them securely. Tagged objects were then dropped into public garbage bins, brought to recycling centers, returned to retailers via take-back programs, dropped on the curbside of residential homes, or, in a single case, left provocatively in public space. We quickly realized that additional measures were necessary in order to protect the tags from physical damage, ensure proper signal transmission, and conceal them to prevent manual removal. In order to achieve these partially conflicting goals, we experimented with different methods for protecting the sensitive devices.

This turned out to be an assembly problem of a very peculiar kind, since we needed to be able to secure a tag to a wide range of materials and geometries. Encasings of latex mold rubber and carbon fiber provided effective protection, but were too complicated and time-consuming to produce on the site of deployment. We settled on a process which covered the sealed tag with a 1-2 inch thick shock-absorbing layer of sturdy foam based on epoxy resin, a quick-setting material also used for insulating and patching boat hulls. For some electronic waste, tags could also be embedded into the interior circuitry of the object. Special care was given to devices with metal cases, to prevent Faraday cages from blocking of blocking wireless signals.

***

The following August, we distributed an additional 600 tags in the city of Seattle, using a threefold approach. A portion of the tags was distributed during a public event at the Seattle Central Library, at which volunteers each brought to us an object to be tagged, which they would then disposed of. For a second portion of tags, our team arranged to visit volunteers at their homes, attaching tags to their prepared garbage objects and having the volunteers dispose of them as they normally would. Finally, the remaining tags were deployed directly by our team across the metropolitan area, covering the full range of materials and garbage types (such as plastic, metal, e-waste, etc.), different geographic areas, and a variety of garbage disposal methods.

We were struck by how strongly the public responded to the project. Within two days of sending out a newsletter announcement in Seattle calling for volunteers, we had received more than 200 responses - far beyond our expectations. It suggested just how engaged the general public could become in researching where trash goes. This also contributed to our decision to set up subsequent deployments of many tags with expanded involvement by local volunteers. Following our August experiment in Seattle, two smaller experiments were carried out in the cities of New York City and London.

In September of 2009, preliminary results from these efforts were shown to the public in exhibitions at the Seattle Public Library and the Architectural League of New York. In both exhibitions, a real-time visualization based on collected sensor data provided a compelling portrait of the journey of the tracked garbage, item by item. Each representation included a photo with the description of the disposed object, where and when it was thrown away, and an animated view of the object's trajectory though the waste system. These data visualizations were combined with a video composed by German artist Armin Linke for this exhibition, which showed scenes from the inside of waste management facilities, juxtaposing clean abstract data with the brute physicality of waste.

***

In October 2009, building on experiences from this initial phase and the exhibitions, we prepared for our largest deployment to date. In order to launch 2,200 location sensors in the metropolitan area of Seattle, we extended our collaboration with local citizens. We recruited volunteers through an open call in local media, in which we asked them to provide garbage items from their own households according to a wish list of items. From the pool of volunteers who registered through our web site, we selected around 100 homes to visit, in order to achieve an even geographic distribution of disposal points across the whole region.

During the following three weeks, four mobile deployment teams visited the homes of selected volunteers, elementary and high schools, and private institutions. The Seattle Central Library served as a base of operations for preparing the tracking devices and the materials necessary for the deployment, as well as for training lead volunteers who joined us during visits to homes. On site, the team attached tags to trash using layers of protective epoxy foam, and documented each item, its material properties, and the time and location of its disposal.

In order to achieve a diverse and representative selection of household wastes, we prepared a 'wish list' and some guidelines for what we were looking for. Two main factors shaped the selection of trash for tagging:

— Primarily, the list was based on the taxonomy used by the Environmental Protection Agency to divide municipal solid waste into categories based on contained materials (Office of Solid Waste 2008b). These material categories include organics, paper and paperboard, glass, plastics, metals, and rubber, leather and textiles.

—The second important influence in selecting products was the nature of the waste collection system in Seattle. In this system, different mechanisms capture different types of waste; for example, single-stream curbside recycling collects aluminum cans, while food and yard waste collection handles organic scraps, and fluorescent light bulbs are collected at household hazardous waste centers (Seattle Public Utilities n.d.).

Extra consideration was given to emerging waste categories such as discarded cell phones, computers, fluorescent bulbs, and other household hazardous waste, which are increasingly prevalent in the waste stream (Office of Solid Waste 2008a). These emerging waste sources have an array of disposal mechanisms, many of which are provided by private sources, such as manufacturer take-backs, store drop-offs, and mail-in programs. The collection rates through these mechanisms, as well as their environmental impacts and trade-offs, are not as established as those of municipal solid waste.

***

While we varied and refined our procedures throughout the project, each of these deployments actively involved local citizens. Besides creating public awareness about the project, its goals, and the questions it raises, this also ensured that we could distribute our tags in a way that closely matched how people in that city actually disposed of their trash. Recruited volunteers came from local neighborhoods, schools, public libraries, institutions, and companies, collectively providing an incredible array of garbage items. Volunteers also joined us in attaching sensors to waste items and documenting the process. Some drove us in their private cars to deployment locations across the city. Finally, volunteers were instrumental in spreading the word and helping to recruit others for the experiment. Most volunteers who signed up for the experiment had a strong interest in environmental issues and technology. When asked for their reasons for participating, they expressed curiosity and a lack of information about the waste management process.

Results

I hadn't thought about the trash having multiple stops between me and a landfill. I also realized I have no idea where my local landfill is.
—Trash Track volunteer survey

***

The Trash Track deployments in Seattle produced multitudes of data; each electronic tag regularly reported its location up to the point of its destruction or permanent loss of signal. Thus, the data could be visualized as traces in space and time, following each trash item's path from its disposal point to its last known location. Coupled with descriptions and pictures of the trash, we were able to show how different types of waste traveled across the country.

However, these traces alone could not tell us about why trash traveled the way it did, nor help us understand what were the functions of each resting point or final destination. We had to supplement our data with thorough analysis using maps, satellite images, published information, and direct contact with waste processing facilities. Performing this manual process for each tracked item helped us determine where it had gone, its mode of transportation, and whether it ended up somewhere it should not have.

***

In this last section we will discuss two typical traces of waste products that were tracked during the experiment: a printer cartridge and a lithium rechargeable battery. Each item's trace illustrates the complexities and uncertainties inherent in predicting the movement of trash.

The trace of the printer cartridge (Figure 1) reveals one way that waste removal chains operate on a national scale. The direct path from Washington to Tennessee, spanning only six hours with no reports along the way, implied this item was shipped by air freight.

We saw from closer inspection that the volunteer disposed of the printer cartridge from a residential neighborhood in the north part of Seattle on Saturday, October 24th, as shown in Figure 2. However, exactly how they disposed of it was unclear; after leaving the house, the tag reported from a street three blocks away, but within the same neighborhood, implying that the cartridge was in transit by car or truck. Because the tag never arrived at a waste transfer station, we assumed that the volunteer personally dropped off the item at a special drop-off point, possibly a retail store collecting recyclable print cartridges.

The next stop of the printer cartridge was a large truck station as shown in Figure 3. Because we could not identify any specialized waste removal facilities nearby, we concluded this was a logistics hub and transfer point between freight trucks.

The cartridge then traveled to Tacoma International Airport, where it arrived at 3:40 am on the morning of October 25th (Figure 4). The GPS coordinates match with the location of a FedEx air freight terminal within the airport.

From there, the tagged printer cartridge flew to Memphis, Tennessee. The reported GPS coordinates indicate that the cartridge arrived at the FedEx air freight terminal at Memphis International Airport by 9:40am on October 25th (Figure 5), no more than six hours after having arrived in Seattle-Tacoma. Incidentally, Memphis Airport is the FedEx Express 'Super-Hub,' processing a large portion of the packages shipped by the company.

The tagged printer cartridge continued its way by air to Nashville International Airport in Tennessee, arriving the same evening. It sat for two days in a US Postal Service facility nearby.

At last, the printer cartridge completed its journey at a facility in La Vergne, Tennessee (Figure 6). The EPA Facility Registry System identifies this location as an e-waste recycling center called SimsRecycling, which recycles, recovers, and remanufactures electronic waste and office products. It is surrounded by industrial buildings but also borders several suburban residential communities.

The ultimate fate of the printer cartridge remains unknown. Since this was the last location report received from this object, it is very likely that the location sensor was destroyed or separated upon entering the facility. The printer

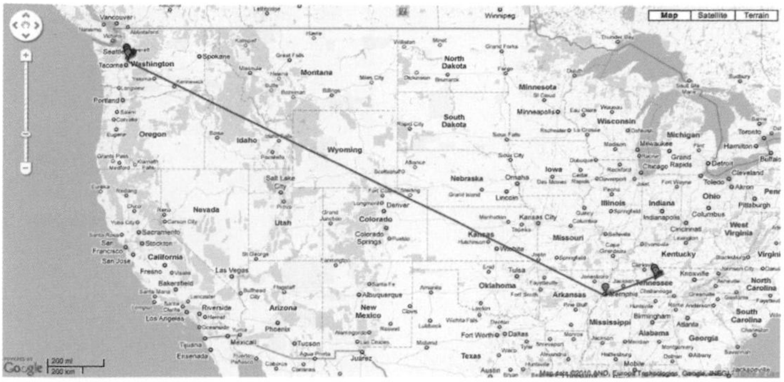

1. Trace of printer cartridge deployed from a residential household in Seattle, WA to an E-waste recycling facility (SimsRecycling) in La Vergne, TN.

2. Location of disposal for the printer cartridge.

3. Cartridge transported to a freight truck facility.

4. Cartridge arriving at FedEx facility in Seattle-Tacoma International Airport.

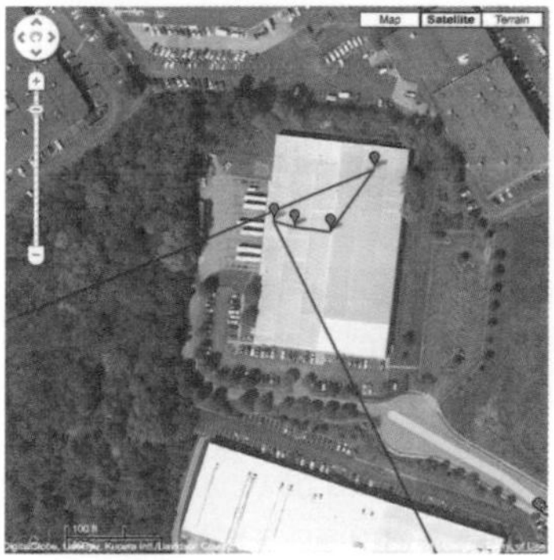

5. Cartridge arriving at Memphis International Airport.

6. Cartridge arriving at SimsRecycling, an e-waste recycling facility in La Vergne, TN.

cartridge travelled 3,210 miles from its disposal location to reach this point, by a combination of road, rail, and air freight.

***

By contrast, the trajectory of a tagged lithium rechargeable battery showed much more uniform motion across the US, as it travelled east towards the Twin Cities (Figure 7).

Closely examining the origin point of the battery shows it starting from a Seattle post office on the afternoon of October 29, as shown in Figure 8. We assumed that the user chose to dispose of the battery through a mail-back program for recycling hazardous batteries.

The lithium battery appeared to travel south by truck to Federal Way, Washington, where it arrived and stayed for the evening in another US Postal Service facility, as shown in Figure 9. On the morning of October 30, it left this facility to travel eastward, reporting several times from I-90.

The battery steadily rolled east, through Idaho, Montana, Wyoming, South Dakota, and

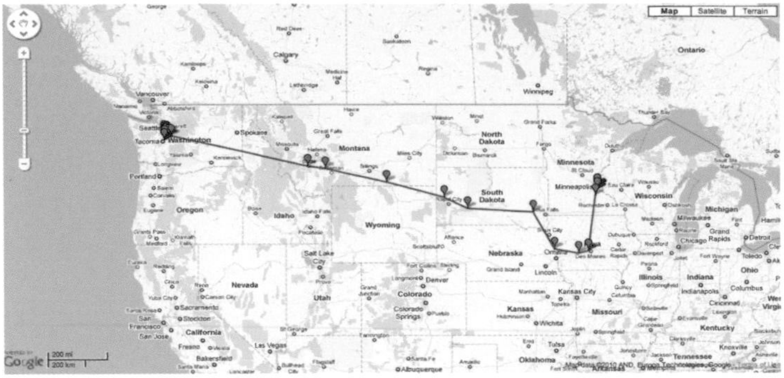

7. Small lithium battery trucked from Seattle, WA to a mecury recovery facility in Roseville, Minneapolis.

8. Lithium battery starting at a U.S. post office.

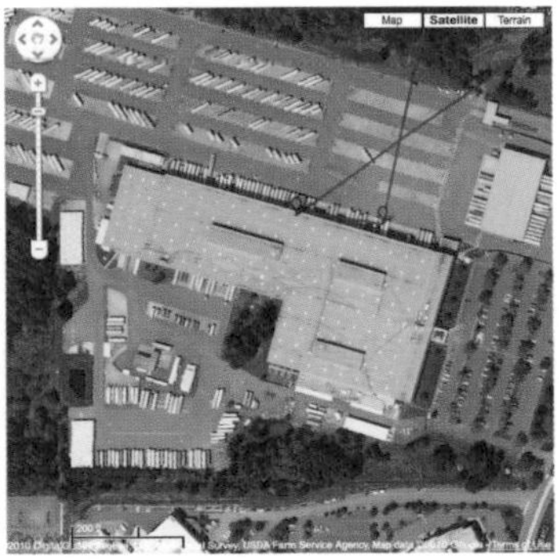

9. Lithium battery leaving transportation facility toward highway 90.

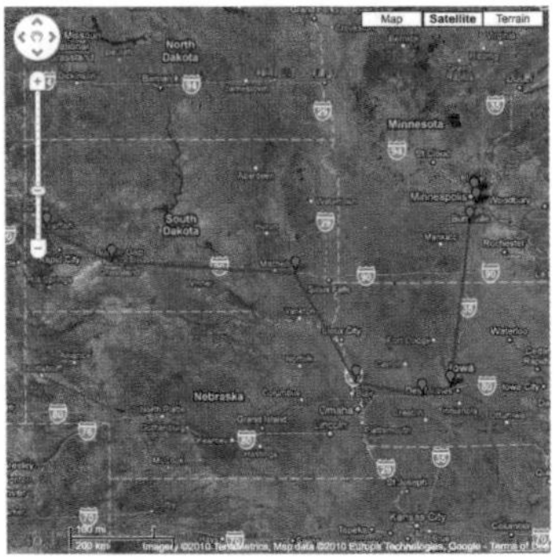

10. Lithium battery transported on highway 90, highway 29, highway 80 and then on highway 35.

11. Lithium battery end-of-life destination: Mercury Waste Systems.

Iowa, always reporting from interstate highways. There were some missing reports, which could have resulted from the tag temporarily losing signal, either from being buried deep within the truck or entering an area without cell phone reception. In northwest Des Moines, the battery arrived at yet another US Postal Service building. It stayed for several hours before heading north to Minneapolis on the evening of November 1, as shown in Figure 10. Travelling at full speed along I-35, it arrived at an Eagan post office on November 2.

Finally, on November 3, the lithium battery arrived at a hazardous waste recovery facility called Mercury Waste Solutions in Roseville, Minnesota. (Figure 11) Mercury Waste Solutions partners with Waste Management to run its LampTracker program, confirming that the volunteer used a special LampTracker container to dispose of the battery. The sensor sent its final reports from the facility; we concluded that the battery and tag were likely separated here, and that this was the likely end of the battery's 3000-mile journey.

These accounts illustrate the rich traces that can be gleaned from tracking two household hazardous waste items, and how their meaning is subject to interpretation. Geographic coordinates alone lack context; their meaning has to be inferred by making educated guesses, comparing the reported locations with those of known facilities, and by deducing transportation modes from movement patterns in time and space.

When combined with the physical characteristics of the object being tracked, these traces can be used to estimate the ecological impacts of the transportation process, such as the amount of carbon dioxide emitted or energy used in transportation and disposal. Yet, these accounts also show how difficult it is to draw conclusions from anecdotal traces, since many events in the removal process, such as how efficiently trucks and train cars were packed with transported waste, remain unknowable.

Reflections

[Trash Track] made me more aware of details on how to dispose of certain items. Before, I had been too lazy to look up where to take an old laptop or light bulbs. I didn't want to throw

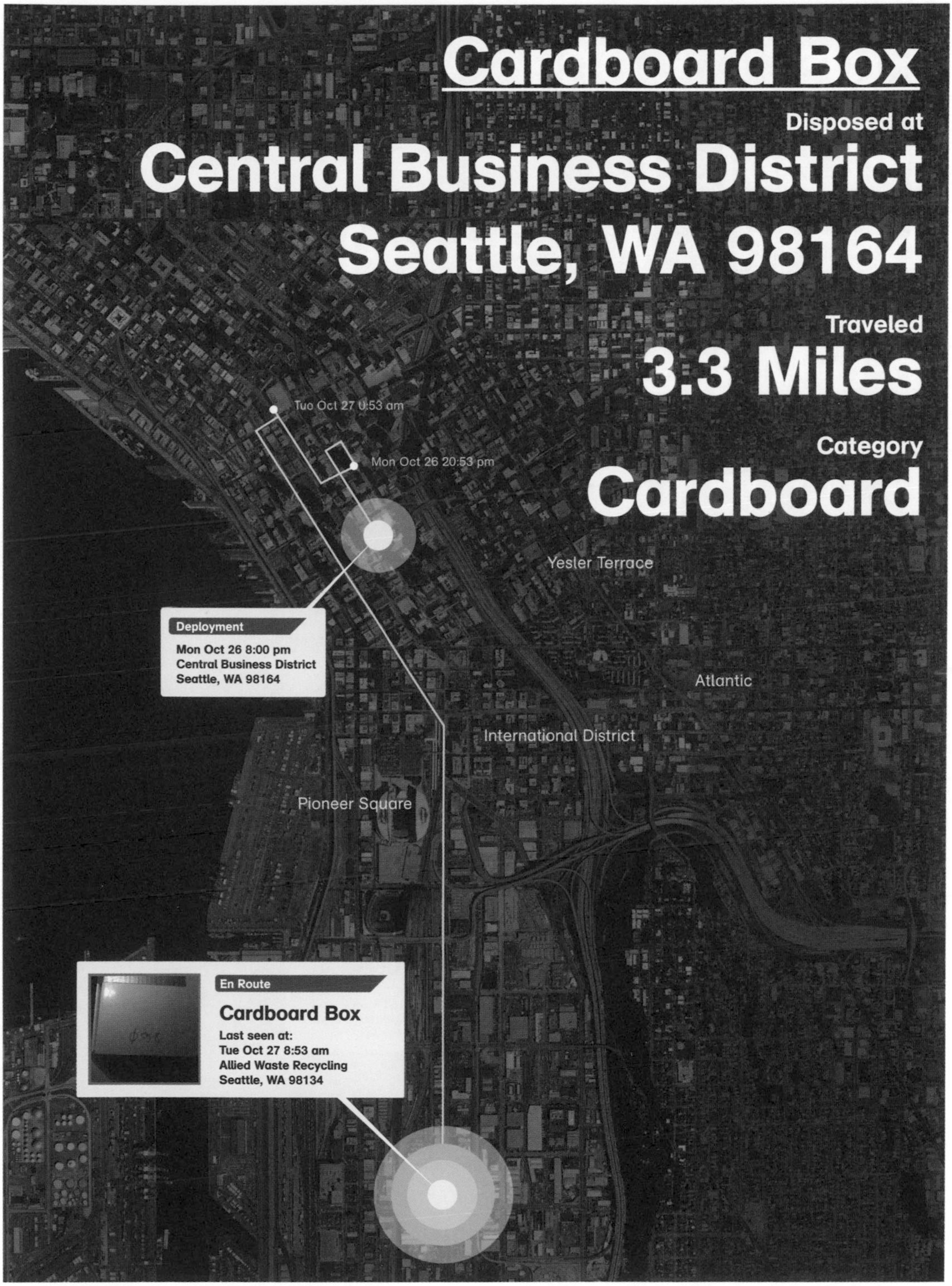

The recorded trajectory of a piece of cardboard on its way to the material recovery facility in South Seattle. Image by E Roon Kang and Eugene Lee.

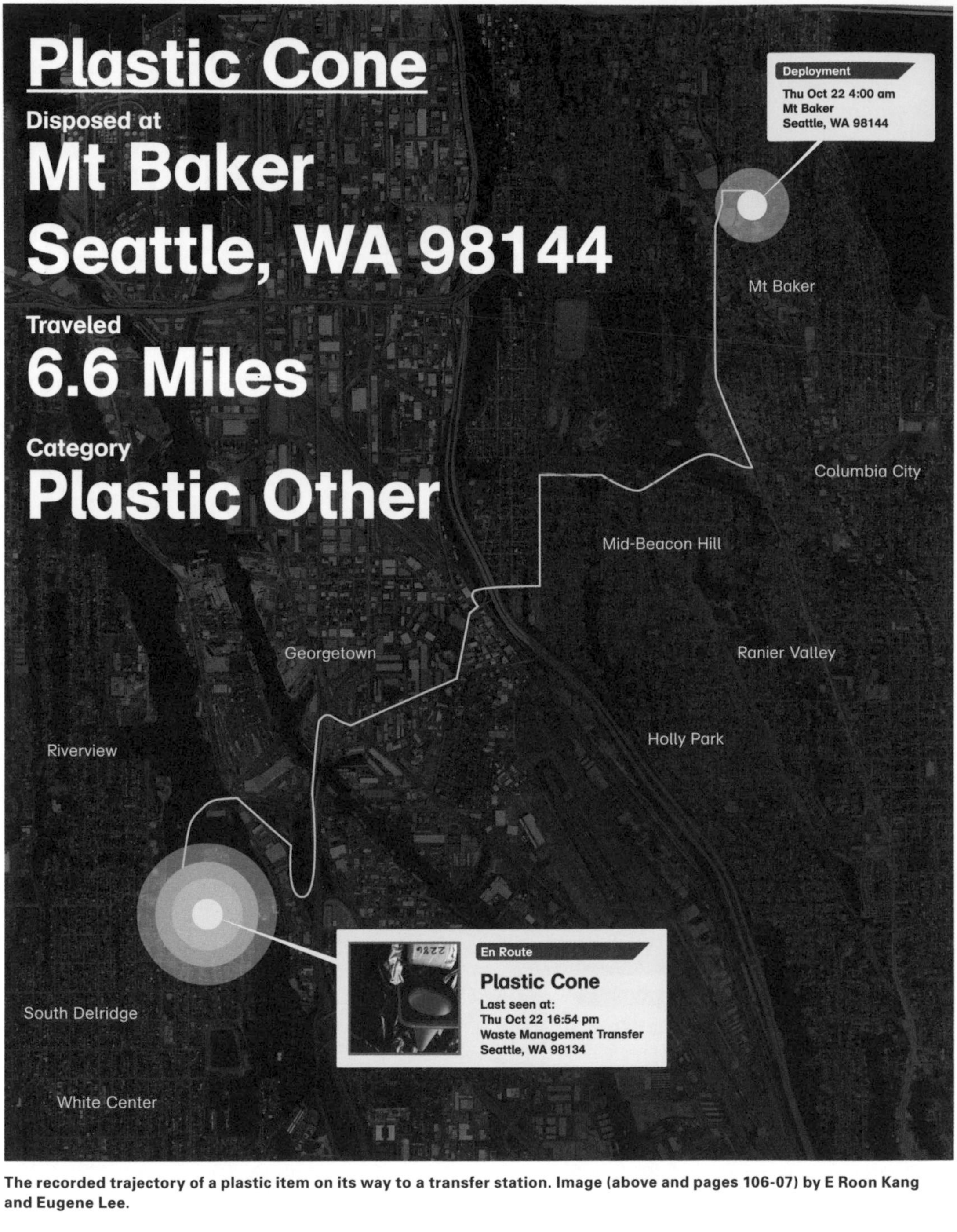

The recorded trajectory of a plastic item on its way to a transfer station. Image (above and pages 106-07) by E Roon Kang and Eugene Lee.

them away, so they just sat around. Trash Track made me motivated to learn the proper disposal process for such items.
— Trash Track volunteer survey

***

One of the most surprising things we discovered was how strongly the project resonated with volunteers, peer researchers and the general public. Despite the bureaucratic aura of public services such as waste removal, trash turned out to be a very emotional topic. The people we met in the course of the project deeply cared about the fate of garbage and recyclables, and wanted to find out how the city deals with them.

Trash Track was originally developed for an art exhibition, and the final project shared many similarities with a participatory art piece. Similar to other projects by the lab that were developed around a public exhibition, Trash Track illustrates SENSEable City Lab's design, action, and intervention-oriented vision for research. Exhibitions, internally called 'urban demos,' play an important role in the work of the lab, not only for discourse with the general public, but also as a nucleus around which we can set up further research. The urban demo is a broad vision of research, boiled down to an actual intervention in urban space, that communicates a vision, demonstrates a possible technical implementation, and provides an approach for producing data that can be followed up by later scientific exploration.

In that sense, it is more than a traditional scientific experiment under controlled conditions; its purpose is not only to collect data, but also to facilitate discourse. Trash Track is an urban demo in the sense that it shows a glimpse of what the future city could look like. Even before going into the details of a specific trace, visitors and volunteers were confronted with the complexity of the waste removal system, knowledge that might change their attitudes and behavior as consumers.

In that context, one interesting question is what role the museum or library could play in such a discourse. According to Jodee Fenton, Fine and Performing Arts Coordinator at the Seattle Public Library, they see their role as engaging the community around relevant topics by providing the best information available, without endorsing a specific view or goal. While the project created a challenge for their infrastructure and institutional culture (imagine volunteers bringing heaps of trash and researchers pouring epoxy foam in the library's pristine, Rem Koolhaas-designed lobby), it still aligned well with the public discourse they sought to promote.

Our experiences with the exhibition validated some assumptions we mentioned earlier — in particular, the assumption that access to raw, personalized datasets can have a higher value for the citizens than traditional forms of data representation. Volunteers developed a strong sense of ownership or emotional attachment to the data generated by their donated item — a fact that we acknowledged during the third deployment by creating a website for volunteers, where they could track their own tagged items. They were genuinely interested in the information they helped to generate, and less interested in the public feature of their names or photos of their donated objects. But the meaning of the data can be perceived on different levels. While the explicit information acquired through the experiment, the destinations and trajectories of waste, facilitate insight into the system, the data is also meaningful on a more basic level. The complexity and the patterns of the whole system suggested by the data, or the mere movement of waste, are meaningful pieces of information for the volunteers.

***

The Handbook of Solid Waste Management identifies a number of issues the field is struggling with, many of which stemming from a lack of information. This starts with the need for common definitions of how to categorize waste. The biggest issue, however, is the lack of quality data about all aspects of waste removal. As a result, it is difficult to answer questions of how much waste is generated, since not all of that waste is properly reported. Experts also stress the current lack of consistent, predictable enforcement of regulations, which also relies on sound data and monitoring, particularly on transfers between states and countries (Kreith and Tchobanoglous 2002).

Trash Track substantially contributes to the availability of quality information about the waste removal system, providing the basis for an integrated treatment of waste, from generation to collection, processing, and recycling or disposal. The project pursued three important goals: to develop appropriate tracking technology given the real-world conditions of waste removal; to generate a bottom-up

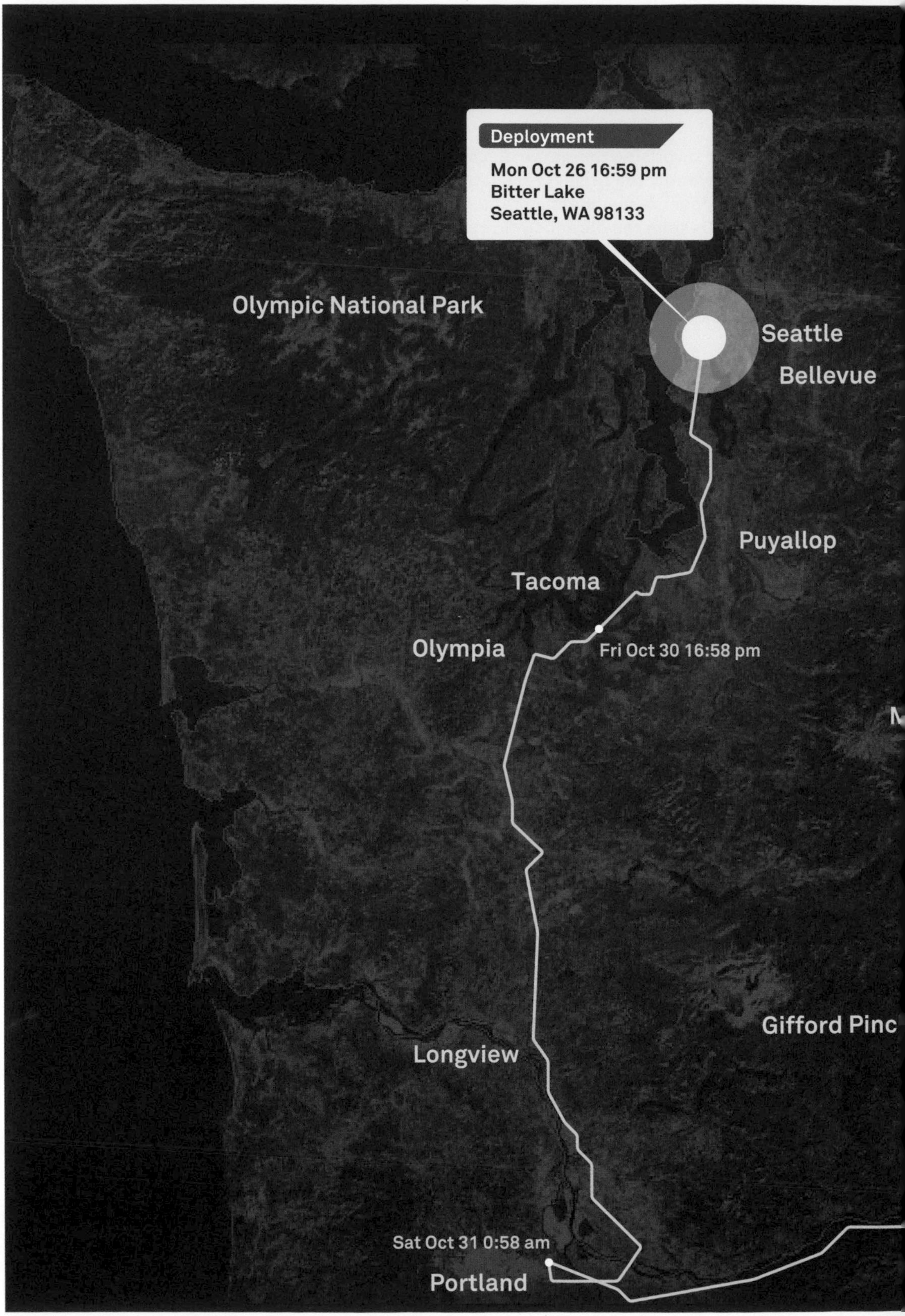
Deployment
Mon Oct 26 16:59 pm
Bitter Lake
Seattle, WA 98133
Olympic National Park
Seattle
Bellevue
Puyallop
Tacoma
Olympia
Fri Oct 30 16:58 pm
Gifford Pinc
Longview
Sat Oct 31 0:58 am
Portland

Old Sneaker
Disposed at
Bitter Lake
Seattle, WA 98133
Traveled
337 Miles
Category
Shoe
ier National Park
onal Forest
En Route
Old Sneaker
Last seen at:
Mon Nov 2 8:35 am
Columbia Ridge landfill
Arlington, OR 97812

view of the removal chain that advances the understanding of its processes and potential weaknesses; and to raise public awareness by illuminating a process that we rely on greatly, but understand poorly.

On all three levels, the project made substantial contributions. On the technical end, we developed a tracking tag that can survive for extended periods of time in waste streams, and can operate on a global scale using infrastructure that is available everywhere. In several iterations, we developed a packaging strategy that protected the sensors and kept them firmly attached to tracked waste. Trash Track also generated a data set that presents a unique, integrated view of the system, its full reach, and the companies involved. Finally, judging by the feedback we received from the public, the project created a vivid representation of a process that is usually hidden and unobservable. We hope that future researchers and institutions can build on our experiences and use Trash Track to advance our understanding of the city and its improvement.

Overall, the project targets a shortcoming in the waste removal system by generating information that has not been available before. It allows the potential for investigating the fate of individual discarded items and touches on issues related to transportation logistics, volume and waste movement. While this information helps us to better understand how the network can be managed, a key aspect remains the project's impact on the public, the feedback to the private individual. If we know where our trash goes and how long it takes to get there, will it have an impact on our future production of waste?

ACKNOWLEDGEMENTS

Waste Management was the main partner and sponsor of Trash Track throughout all phases of the project. Seattle Public Utilities, Qualcomm, and Sprint provided close technical support in setting up and running the experiments. Both the Architectural League of New York, who originally commissioned the project, and The Seattle Public Library hosted exhibitions of Trash Track results. The Seattle Central Library was a crucial home base for all of our activities in Seattle.

NOTES

1. Bay Area Rapid Transit, "BART - For Developers," 2008, http://www.bart.gov/schedules/developers/index.aspx.
2. New York City Government, "311 Online," 2010, http://www.nyc.gov/apps/311/.
3. O. Okolloh, "Ushahidi, or 'testimony': Web 2.0 tools for crowdsourcing crisis information," Change at Hand: Web 2.0 for Development (2009): 65.
4. Mark Weiser, "The computer for the 21st century," Scientific American (September 1991): 94-104.
5. J. Bleecker, Why Things Matter: A manifesto for networked objects–cohabiting with pigeons, arphids and Aibos in the Internet of Things (Near Future Laboratory, http://www. nearfuturelaboratory. com/files/WhyThingsMatter. pdf, 2005).
6. Katharina Kummer, International management of hazardous wastes: the Basel Convention and related legal rules (Oxford University Press, 1999).
7. Greenpeace International, "Following the e-waste trail - UK to Nigeria," 2008, http://www.greenpeace.org/international/photosvideos/greenpeace-photo-essays/following-the-e-waste-trail.
8. Curtis C. Ebbesmeyer et al., "Tub Toys Orbit the Pacific Subarctic Gyre," Eos 88, no. 1 (January 2, 2007): TRANSACTIONS AMERICAN GEOPHYSICAL UNION.
9. Eric Paulos and Tom Jenkins, "Urban probes: encountering our emerging urban atmospheres," in Proceedings of the SIGCHI conference on Human factors in computing systems (Portland, Oregon, USA: ACM, 2005), 341-350, http://portal.acm.org/citation.cfm?id=1054972.1055020.
10. Bruce Sterling, Shaping Things (MIT Press, 2005).
11. Seattle Public Utilities, "Seattle Public Utilities -- Services," n.d., http://www.cityofseattle.net/util/Services/index.asp.
12. Office of Solid Waste, Electronics Waste Management in the United States: Approach 1, Final (United States Environmental Protection Agency, July 2008).
13. Frank Kreith and George Tchobanoglous, Handbook of solid waste management (McGraw-Hill Professional, 2002).

REFERENCES

— Bay Area Rapid Transit. 2008. BART - For Developers; http://www.bart.gov/schedules/developers/index.aspx.

— Bleecker, J. 2005. Why Things Matter: A manifesto for networked objects–cohabiting with pigeons, arphids and Aibos in the Internet of Things. Near Future Laboratory, http://www. nearfuturelaboratory. com/files/WhyThingsMatter. pdf.

— Ebbesmeyer, Curtis C., W. James Ingraham, Thomas C. Royer, and Chester E. Grosch. 2007. Tub Toys Orbit the Pacific Subarctic Gyre. Eos 88, no. 1 (January 2): TRANSACTIONS amERICAN GEOPHYSICAL UNION. doi:10.1029/2007EO010001.

— Greenpeace International. 2008. Following the e-waste trail - UK to Nigeria. http://www.greenpeace.org/international/photosvideos/greenpeace-photo-essays/following-the-e-waste-trail.

— Kreith, Frank, and George Tchobanoglous. 2002. Handbook of solid waste management. McGraw-Hill Professional.
New York City Government. 2010. 311 Online. http://www.nyc.gov/apps/311/.

— Office of Solid Waste. 2008a. Electronics Waste Management in the United States: Approach 1, Final. United States Environmental Protection Agency, July.
———. 2008b. Municipal Solid Waste in the United State: 2007 Facts and Figures. United States Environmental Protection Agency, November.
— Okolloh, O. 2009. Ushahidi, or 'testimony': Web 2.0 tools for crowdsourcing crisis information. Change at Hand: Web 2.0 for Development: 65.
— Paulos, Eric, and Tom Jenkins. 2005. Urban probes: encountering our emerging urban atmospheres. In Proceedings of the SIGCHI conference on Human factors in computing systems, 341-350. Portland, Oregon, USA: ACM. doi:10.1145/1054972.1055020. http://portal.acm.org/citation.cfm?id=1054972.1055020.
— Seattle Public Utilities. n.d. Seattle Public Utilities -- Services. http://www.cityofseattle.net/util/Services/index.asp.
— Sterling, Bruce. 2005. Shaping Things. MIT Press.
— Weiser, Mark. 1991. The computer for the 21st century. Scientific American (September): 94-104.

AN INTENTIONAL FAILURE FOR THE NEAR-FUTURE

TOO SMART CITY

DAVID JIMISON AND JOO YOUN PAEK

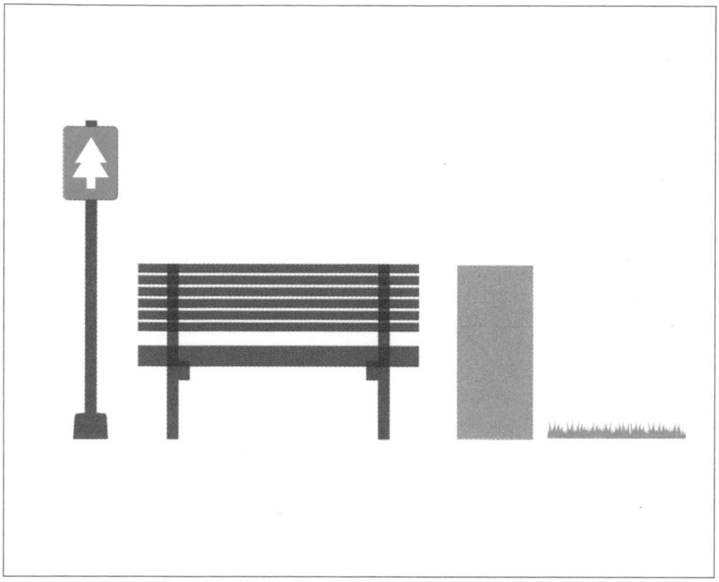

Too Smart City, illustration JooYoun Paek 2009.

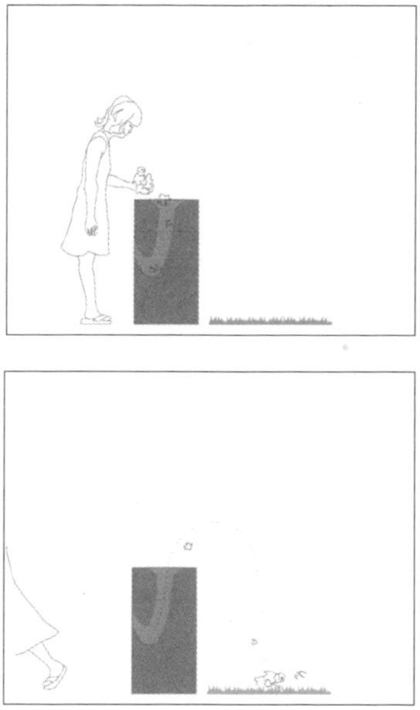

Too Smart Trashcan, illustration JooYoun Paek 2009.

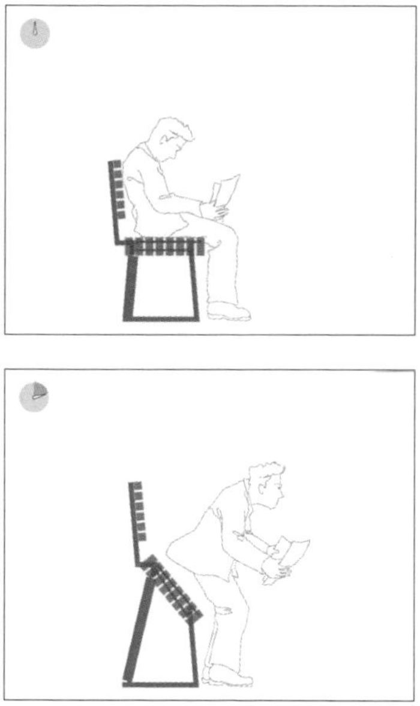

Too Smart Bench, illustration JooYoun Paek 2009.

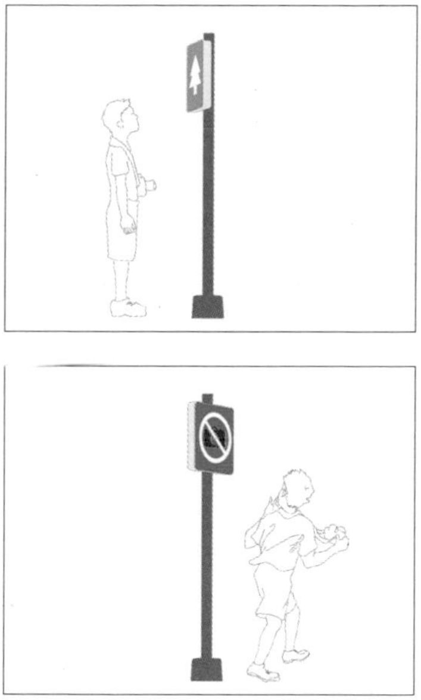

Too Smart Sign, illustration JooYoun Paek 2009.

Too Smart City is a series of public street furniture (a trash can, an informational sign, and a bench) each integrated with artificial intelligence. In each, the over enthusiastic use of technology renders the functionality of the objects nearly useless. The Too Smart Trashcan spits trash back at the guest, the Too Smart Sign turns to face passersby presenting specific rules for them, and the Too Smart Bench ejects vagrants by lifting them up and dumping them off. Together, Too Smart City invokes a city park run haywire. Presented to the public as a representative of the Sentient City in the tradition of the world's fair, and similar technological expos, Too Smart City questions the relationship between urban public spaces and technological progress. Through their failures, the furniture transcend utilitarian function, and become an intervention in the near future.

This chapter introduces Too Smart City, concentrating on the theoretical reasons for engaging with intentional failures, and the applied processes employed to build the devices. The creators, JooYoun Paek and David Jimison, are both artists and technologists, hybrids of two communities which each have specific

Too Smart City gallery installation, photography JooYoun Paek 2009.

historical precedence in the issues surrounding the Sentient City. Pulling from these different traditions, the pair engages in scientific user tests, delves into urban design studies, and works with the latest in technological innovations, to create furniture that is at once timeless and futuristic, marvelous and disastrous.

Too Smart City

One of the most prominent features of Too Smart City is the synthetic, vibrant green grass, on top of which the furniture is placed. The grass leads out from the gallery wall in an organic curvy shape, similar to the mound of a small hill. Overall, it measures approximately eight by sixteen feet in size. The grass is an important feature to the piece, because it establishes the three furniture as coexisting within a space, that of a park. The grass serves to differentiate the furniture from the rest of the gallery, and maintains the simulation of the park. In addition to the visual impression it holds, the grass feels soft under the feet of the guests, helping to further the illusion of a park.

In the foreground of the furniture, the Too Smart Sign stands at six feet tall, a polished aluminum structure with a black plastic housing around a screen. The rectangular base of the aluminum housing is an exact duplicate of the bases found on street lights in many cities. Along the sides of the base are long horizontal slots, fanned outward, a design common in public signs. Were it not for the black screen mounted at the top of the sign, the entire structure would appear ordinary in any urban setting.

The monitor was selected for its 2:3 ratio, uncommon for computer monitors, but typical of street signs. The display alternates between a number of park illustrations from "No Smoking" to "No Standing." Each illustration is predominantly iconographic and all are from the universal subset of signs commonly found in parks. On top of the monitor there is a small gray sonar sensor. The sonar sensor is mounted directly in the middle of the screen and tilted slightly down so that it is able to detect the movement of people walking by.

Too Smart Sign details, photography Jena Gagliano 2009.

The top of the sign rotates on its axis 270 degrees over the course of two minutes. As it rotates, the sonar sensor compares what objects are near it to those of its initial readings of the gallery space. Upon completing a survey of the entire arc, the sign computes if any guests are standing closest, and quickly spins on its axis to face them. As it rotates, the sign switches its message, using a new, personalized message for the guest. The sign remains locked into position, facing the guest. As the guest moves around the sign, the sign rotates to continually face them. Only after retreating from their location, and moving away from the sign, does the sign stop following them and return to its previous state of rotating back and forth on its axis.

The Too Smart Trashcan is a forty-gallon polished aluminum container, which, aside from a black electrical cord snaking out of its base, appears rather commonplace. Mounted on the wall next to the trash can is a clear plastic shelving unit containing ten separate drawers, and in each of the drawers, there is a different type of trash. The center of the trashcan bellows inward creating a sliding surface for trash to move into the seven inch entrance. If one were to look carefully into this entrance, one would see a red laser beam at approximately four inches down, shooting horizontally across the hole.

Placing a piece of trash into the receiving hole breaks the contact between the laser beam and the light sensor across from it, thus notifying the Too Smart Trashcan of new trash being received. The trash rests on a platform eight inches below the top of the trash can

Too Smart Trashcan and trash samples, photography Jena Gagliano 2009.

Too Smart Trashcan: laser beam inside to detect trash coming in, photography Jena Gagliano 2009.

Too Smart Trashcan: trash jumping out of the trashcan, photography Jena Gagliano 2009.

entrance. Under this platform is a metal detector which is connected to the trash computer circuitry. When new trash is received, this metal detector compares the amount of metal currently near it with its default amount. Bear in mind that the entire housing of the trashcan is made of aluminum, so the variation between can and not is rather minor. For this reason, the range being used by the metal detector is narrow and thin, so that only trash resting on top of the platform is detected.

If the trash contains over a certain threshold of metal, it is deemed worthy of being recycled. In this case, the platform lowers 90 degrees, allowing the trash to pass into the bottom bin of the trashcan. If the trash is determined to be unacceptable, a high powered fan is triggered from a side chamber within the trashcan body. This fan blows wind at a velocity of 80 miles per hour, causing the trash to be blown out. The strength of the fan was calibrated so that the average piece of trash would float about three feet above the trashcan before flying to the side. The ventilation system itself is angled in a way so that the trash first flies upward and then lands in a specific direction each time. This system was used to ensure that guests would be able to react to the trash flying towards them, and also to keep rejected trash in one specified area of the Too Smart City area.

The Too Smart Bench is a gorgeous two person seater made of polished aluminum and stained oak wood. The seat of the bench rests at two and half feet from the ground, and the back stands at another foot in height. Upon closer inspection, there are a series of small light sensors, embedded into square aluminum posts, which are lined up between the gaps in the wood on the bench seat. These sensors are spaced at six inches apart horizontally, and four inches apart vertically. This distance was selected to ensure a minimum of four sensors would be covered when the average sized posterior was seated on top of them.

When a user sits on the bench, the sensors match the pattern of those covered, to those that are not, determining whether the change in light is universal, or in the shape of a person's posterior. After resolving that it is indeed a human posterior, the embedded computer notes the time and position of the guest. The amount of space they take while sitting corresponds to the amount of time they are allotted upon the bench. When this time has passed, the two actuators in the back legs of the bench are activated, causing the bench to lift up 18 inches. This motion creates an incline which shifts the guests weight from being across the entire base of the seat, and onto their knees. Through this method, the bench essentially stands the person up on their feet. After reaching its raised capacity, the bench waits for fifteen seconds after detecting that users are no longer seated in that position, and then lowers the back legs to return to its normal state.

The Too Smart City furniture are intended as an individual art work, together forming a specific installation evocative of a park. Were these

Too Smart Bench, photography Jena Gagliano 2009.

Too Smart Bench in motion, photography JooYoun Paek 2009.

furniture viewed on their own, their artistic message would center on each of their specific functions. By placing the works together, we are generating a collective message about the Sentient City. Parks are ubiquitous forms of urban space, and are indicative of the larger city. By extension of the park, we are attempting to imply characteristics of the larger Sentient City.

The Sentient City as Science Fiction

We approach the Sentient City from the perception that it does not yet exist other than as a concept. For a city to be sentient, one assumes the ability to feel and think. There are, of course, no cities that feel and think in the human sense of the term. By extending the meaning into the computational realm we find that cities do sense activities through the assorted technologies embedded in them. Certainly, security cameras, networked throughout a city, provide a form of sensing, a method of attaining awareness of what is occurring in the city's surroundings. While such sensors monitor change, a sentient city would be able to comprehend such changes over time. There do exist large data systems which enable the persistence of memory across events. However, these technologies, if they do qualify as sentience, do so only as a very elementary mode of such urban environments.

Overall, the notion of the Sentient City is often concerned with a city that has gained consciousness, and in these dialogs of the sentient city, there is an element of anthropomorphism. To us, the sentient city is thus both the actual technological progression towards a new computationally enabled urbanism, and the mythological anthropomorphized city as living organism. We define this conceptual domain of the sentient city as science fiction, both in relation to the literary genre, with its wild and fanciful cities of the future, but also in the manner that novel technologies generate new ideas about the upcoming future.

The cities of sci-fi works such as Star Trek and Blade Runner are in actuality backdrops to dramatic plots, serving more for story arc than as prototype for any actual future cities. However, these sci-fi cities create a legacy, or perhaps a contemporary mythology, of what one can expect in the future. Even though many of these notions of the Sentient City are not statistical likely, they remain difficult to resist. After all, this is the vision of the future city that we grew up.

In contrast to sci-fi, scientific research is grounded in realism, and is a slow progression. However, scientific and technological institutions generate their own form of fiction, a hype around their developments. This often appears in the form of the scenario these institutions use to explain the potential of a new product or development to the layperson. These stories often promote best case scenarios of the devices, and are at times based more on hopeful projections than actual capabilities. This layer of exaggeration is more prominent in technologies which attract public attention, and is generated by miscommunications from scientists to public relations to journalists and finally to the public. This type of science fiction is apparent in discussions of the sentient city. While components of such a city are coming increasingly into existence, the vast infrastructural requirements to generate a truly sentient city are still very much nonexistent.

In researching the Sentient City for the current project, we were more interested in the popular culture myths about the future city then any specific implementations of one. While there continue to be advances in technologies, and the urban landscape will evolve with ever increasing computational capabilities, the trajectory of such developments is informed, at least partially, by cultural expectations. As artists whose medium is technology, we find the greatest impact by using our technological innovations to intervene in the cultural conversation about the near future.

Too Smart City is therefore built as a fictional space of its own, generating new stories, albeit ones of computational failure. It is in part the inheritor of a strong tradition in the promotion of science, such as Tomorrowland, which Walt Disney promoted by saying "The Tomorrowland attractions have been designed to give you an opportunity to participate in adventures that are a living blueprint of our future." Like Disney, we are creating a future world in miniature, an interactive installation, that lets guests fully engage with what is possible.

In order to create the illusion of Too Smart City being futuristic, we needed to use technologies that behaved in a manner expected of a future city. Thus, despite designing a simulation, we are still required to build advanced sensing and computational technologies. While, on the one hand, we are creating a fictional space steeped in traditions of experience design, on the other, we are researching and developing new technologies that are part of the near future.

Selecting Technologies for the Near Future

In using the term near future, we are referring to the approaching time period of five to fifteen years from now. The near future is fascinating to us because it is close enough that we can accurately predict some scenarios, but far enough away that it is impossible to be very specific. Instead, the near future is often a slight exaggeration of the contemporary world.

The sentient city relies upon technologies that do not yet exist, but which are nonetheless expected. For us the problem then becomes, how to work together to discuss and critique the Sentient City in a grounded manner. It is a process of starting with research, and then imagining where such technology could end up. To do this, we rely upon works of science fiction in a variety of media as a frame of reference. In addition, we search for new ideas emerging in various technological fields. This research provides a lexicon of specific technologies to which we can refer to in our own ideas. We are thus combining actual technologies with cultural myths to project what one might find within the Sentient City.

We set a further constraint upon our research by limiting the technologies we consider comprising a sentient city to the abilities to sense, to process or think, and to react. The hardware that exists in a Sentient City would include embedded computers, physical sensors, and networked communication technologies. The software to run these systems would maintain awareness of what is occurring and respond to it. However, the Sentient City goes beyond this hardware and software paradigm, in its cultural mythology, which considers it omnipresent and self-conscious.

Between the mythical and the technological, we attempted to select technologies that would simulate the Sentient City. The project demanded both theoretical research and applied labor. This process of both imagining and building simultaneously is vital to our process. The conceptual framework guides the technological development. Meanwhile, the process of coding, wiring, and designing CAD models are all aspects of learning more about the system, which feeds back into our conceptual framework. We must therefore engage our understanding of the Sentient City by building it.

It remains a challenge to create near-future objects with present day technologies. Our process, as artists, is first to imagine a list of devices that we would wish to create and next to select from these based upon those that we feel could be potentially made. Some of our early concepts of the Sentient City were not included in Too Smart City due to the difficulty in building them.

Rather than attempt to invent completely new technologies, where possible, we chose to adapt existing technologies to meet our purposes. For each of our ideas, we attempted to discover methods for using other technologies to create the same desired effects. For example, in Too Smart Trashcan we appropriated toy metal detectors and fans intended for model planes to create the detection and ejection systems. Our strategy was at times more that of a hacker than of the traditional inventor, and much of our time was spent reading over online documentation and running tests to determine the viability of different components.

Aside from tackling the physical hardware, there is also a theoretical question involved in the selection of software methodology. One key factor in this selection was the limitation imposed by using embedded computer chips. In order to keep the embedded electronics to a minimal size, both the Too Smart Bench and the Too Smart Trashcan used an 8-bit microprocessor that could only handle sequential coding.

The other determination for software was in selecting the amount of personification that Too Smart City would have. Artificial intelligence (A.I.) has been around for over forty years, and there is historical precedence for such systems being anthropomorphized. Joseph Weizenbaum's ELIZA is a notable example of such a system. Developed in 1964, the software was able to elicit emotional responses from subjects through clever inversions of their text-based questions. This is one of the first examples of A.I. and establishes a precedent in the field for software systems that exhibit human like intelligence. ELIZA is considered A.I. because it exhibits intelligence, despite being a relatively simple program that does not engage in extended thoughts. Many interactive objects from cars to games elicit a certain level of anthropomorphization, whether intentionally or not. In coding Too Smart City we were aware that people would likely imbue personality onto our objects whether we coded it in or not.

Our theoretical software question then becomes whether our devices should be understood by the public as being sentient. Clearly they should not have over the top, theatrical personalities. Were we to build full animatronic characters, they would upstage the overarching concept of the piece. By tweaking the personality back and forth on the level of performativity, we eventually settled on performances that remained subtle and suggestive, yet allowed for guests to infer the personality of the furniture.

For example, until the final stages of our production, the Too Smart City furniture contained embedded vocal chips enabling each of the units to speak to the guests. "Welcome to the Smart Bench" stated our bench when a guest sat down. "You are now being evicted" it would announce before dumping the user off. In our initial tests we believed this type of performance was necessary to convey that the furniture had sentience. However, once these voices were embedded into the interactive furniture, we found them to be too performative. The talking furniture prompted the guests to reply, assuming the furniture had speech recognition capabilities as well. By removing speech from the devices, guests would still infer personality from their interactions, but without the clichés of talking machines.

Our selections in both hardware and software were based upon creating a simulation of the near future. We crafted our concepts of this from both science fiction and contemporary technologies. After determining the overarching spectrum of possibilities, we set about selecting specific capabilities. While we had created a framework to position our concept of the Sentient City in the larger context of the city, we still needed to determine how sentience would engage with a city.

Sentient City as Space

The question of personality in the sentient city is complicated in part by the fact that all cities have, if not personality, then at least unique character. Each city has a persona that it identifies with, and projects outwardly to the rest of the world. Paris is known as the city of romance, New York City never sleeps, Edinburgh is haunted. Were these cities to gain sentience, we believe it would be as a technological magnifier to the city's persona.

Whatever qualifiers pertain to a city's sentience, it can be assumed that they would be somewhat distinct to their location. Therefore, we do not believe the Sentient City is a homogeneous force, nor that it would be identical across cities. Likewise, the Sentient City would not appear only at one singular location, but would happen across multiple cities. Our implementation of the Sentient City needed to operate between any city, but still retain the specificity of a singular location.

If sentience alone does not generate persona, then it follows that the personality of a Sentient City is acquired from the characteristics of the city that it is in. This phenomenon is similar to the differences between subway systems in cities. The elevated line in

Chicago, the S-Bahn of Berlin, and the BART of San Francisco each take on aspects of their city's personalities. While there is homogeneity in the functional aspects of these subway systems, each transports passengers across a variety of stops within the city for set fees, the characteristics of each are unique to their respective cities. We attempted to extend this correlation between infrastructure and location into our work with the Sentient City.

We are interested in engaging with the Sentient City as a real space, as a real city, transformed into one with sentience. This requires designing with a level of specificity that matches the aesthetics of the area it is in. On the other hand, we want Too Smart City to be a part of the broader discussion about sentience across all urban environments, not just New York City. These two tasks conflict in that one is specific to locale and the other requires abstracting away from each city to discover a more universal element.

Our being from different cities (Paek from Seoul and Jimison from Washington DC) helped us in solidifying our designs for Too Smart City. When imagining how the furniture should look or operate we pictured what variations might occur within the different cities. This concern with regional specificity resulted in selecting design styles that were classic modern, and that can be found in most cities, especially New York City. Our selection of this aesthetic was a best effort solution to a problem that can not be fully resolved. It is an oxymoron to have a location specific style that is internationally universal. Instead, when Too Smart City travels, we plan to modify it as needed to maintain the illusion that such a park could appear in the city in which the piece is being displayed.

Everyday Cities

From our perspective, the Sentient City appears as something rather futuristic, and the notion of inhabiting a city that reacts to what is happening within it, is novel and exciting. Yet one of the facets we are most interested in about the sentient city is in imagining what it will be like after the novelty has worn off, when it is part of the normal, mundane, and everyday life within such a city.

Everyday life is comprised of assorted actions and behaviors performed repeatedly. It fascinates us both due to the large portion of life that it takes up, and also because it is largely ignored. Once new technologies become part of our everyday routines, and we begin to rely upon them, they cease to create the same impression upon us. The cellphones upon which we depend for communication are a good example of this and of technological integration. A similar phenomena would likely occur with the Sentient City. As time passes, the affordances of the Sentient City would become a regular part of our everyday life. It is important to bear this in mind, that even though such technologies are novel and thrilling to us, they will have the greatest impact upon people who take them for granted.

Too Smart City intends to engage its audience in a playful critique of the Sentient City. It is our hope that by interacting with our work, there will be an appreciation for how the city changes, and how it might affect future inhabitants. Were we to present a project that was completely novel, the concern would be that the audience would become transfixed by the novelty of the technology, and not notice the larger issues at hand. While, people often engage in critiques of their cities, these tend towards aspects of structural failure, rather than a deeper imagining of what is possible. Such critiques occur most often when something no longer functions. A common example of this are traffic lights. The public does not debate the installation of a new traffic light system, nor consider their effectiveness in the regulation of traffic. As everyday objects, people react to traffic lights without being conscious of them. It is only when the traffic light breaks, and that reliance upon them ends, that they become noticeable.

With Too Smart City we wanted to invoke this notion of the everyday, to make people aware that these devices are something that people will use on a daily basis. By creating them as everyday objects, we are making implications about the rest of everyday life. If the park bench only allows sitting for a certain time, it stands to reason that life in general is overly regulated.

However, creating the illusion of the mundane is not enough. After it begins to interact with the guest, the furniture would, in many ways, be spectacular. Therefore, the best strategy was to build Too Smart City as furniture which would fail. Each of these failures were specifically related to the everyday expectations of that particular type of furniture. The bench does not allow sitting, the trashcan throws trash back at the user, and the sign overwhelms pedestrians with personal information.

Given that our devices are technological, and not ordinary, presenting them in a way that made them appear to be everyday objects posed a problem. As will be explained later, we had to be very clever in our designs to mask the unusual aspects of each piece so that (at first glance) they would appear common. By selecting a trash can, a bench, and a sign, we could rely upon people not only knowing what these objects were for, but also being so deeply entrenched with what their affordances should be, that they would find it difficult to use them in any other way.

Intervention

Too Smart City engages guests in the possible futures of the sentient city. The project operates as an intervention into a possible future, with each furniture displaying the possible failure of technology. The design of how each piece fails was critical for how the message would come across. There were many potential problems in our execution. We could be too overt in our message, or we could delve too deeply into the technological or urban aspects, thus losing a portion of the audience.

Our choice to use comedy, specifically comedic inversion, enabled us to craft the experiences into a series of stages. In the first stage, the guests encountered the piece, a series of everyday public park furniture on a patch of synthetic grass, in the middle of the art gallery. There is a tension in these ordinary objects being placed in the gallery. They are common place both as art and as technology. As a park, they are inviting to guests to come sit down and relax, an activity that is not customary in the gallery setting.

In the second stage of the experience, guests find the Too Smart City objects transforming from the mundane into the extraordinary. It is at this moment of realization that we intend to be comedic, because it operates as an inversion of the expected behavior. The trashcan throws trash back at the user, the bench forces people to stand, and the sign confronts rather than providing information. Each of the Too Smart City objects fails in direct contrast to their expected functionality.

The third phase occurs after the interaction has ended. Following the experience, each piece returns to its normal state, ready to repeat the process. It is through repetition of the same function that the failure is displayed as being systematic. By executing the problem repeatedly, we are demonstrating that the furniture are intentionally failing, that they are supposed to behave in this manner.

The trash can, bench, and sign not only malfunction, the malfunction is a part of their programming. Since all of the furniture operates in this way, it is implied that these malfunctions are indicative of a problem with the larger Sentient City. The problem is new to these sentient models, and the fault is placed upon poor logic. The trashcan is outrageous because it throws the trash back at the user and onto the ground, covering the nearby vicinity with litter. The desired response is that trashcans were fine prior to this introduction of embedded technologies, and that the technology made things worse.

The use of failure as a strategy is a bit risky, especially since we want the experience to remain enjoyable. First of all, we must build the failure in a way that is not simply frustrating, but rather comedic. This is somewhat assuaged by the pieces being in a gallery, and people not having a need for their actual functionality. Frustration is further eased by having each of the interactions be fun. Being lifted on the bench, catching the trash as it flies back at you, dodging the view of the sign, are all forms of play. This is a fine line, since if the pieces are too fun and enjoyable, the sense of outrage we are attempting to elicit is upstaged by the enjoyment in the experience.

A larger risk when using failure as a strategy, was ensuring that the pieces remained safe for use, that the failure of the pieces could not in any way cause personal injury. For this reason each object was carefully designed to ensure that people could not be hurt. Every area of physical movement had to be designed to avoid pinch points. The sign was built to move on a single axis and at a height that will not knock into other people. The trashcan has a weight threshold, above which it cannot throw the trash out, but drops it into its receptacle. The bench lifts upon its back legs, thus raising the person to a standing position, rather than dropping down and sliding them off the front. For each, failure meant an increased potential for harming people if the designs were not carefully thought out.

The final aspect to consider with intentional failure is that it requires a greater vigilance against unexpected problems. When a work is understood in part because of its not working correctly, any other malfunctions

Mimicking bench (left), trashcan (above) and sign experience (right), Photography ChiungHui Chiu 2008.

can be misread as part of the piece. This unfortunately happened to us, when after a month of solid use, the motor on our bench leg gave out. This actual failure of the bench mechanism confused the message of the rest of our piece. People wondered whether the bench was in fact supposed to work at all. While in hindsight there is little we could do to have averted the problem, it does support the notion that using failure must be done with caution.

Our Process

Our approach in building Too Smart City is largely influenced by our respective training and experiences in interactive media. As a strategy for designing Too Smart City, we chose to combine our more traditional artistic approaches with some of the methods typically employed in the interactive media arena. We borrowed extensively from interactive media communities the processes of user testing and iterative design. This hybrid solution between art and science enabled us to base artistic vision on field testing and research.

It is important to note that there remains a strong tension in producing interdisciplinary work. At times, the strategies of either art or science conflict, and this persists even after the project is completed and the work is received by the public. Each community is able to determine its own aspects within the work, but is not always appreciative of the other aspects. For instance, the scientific community has more trouble understanding the value in creating a piece that eschews functionality for expression. Likewise, the art institution has difficulty in curating pieces in which the gallery attendants must be able to turn knobs and plug devices into their computers. As practitioners in both disciplines, we made our design choices by selecting the arena that the project is best suited for. With Too Smart City, we actively attempt to create a hybrid work that acknowledges being the product of two worlds while not entirely at home in either.

Mimicry

In order to defamiliarize ourselves with our urban environment, we engaged in a process of observation and repetition of everyday behavior in public space, which we term Mimicry. Mimicry is a bit similar to the Situationist International *derives* in the sense that it is an intentional wandering through the city, in order to rediscover it. In Mimicry, we start off with a camera and notebook, traveling across different places in the city and attempting to determine what the affordances of such spaces are.

In areas such as Battery Park we observed the flow of people to and from tourist destinations, bathrooms, water fountains, and informational maps. We were curious how each space seemed defined by a specific activity associated

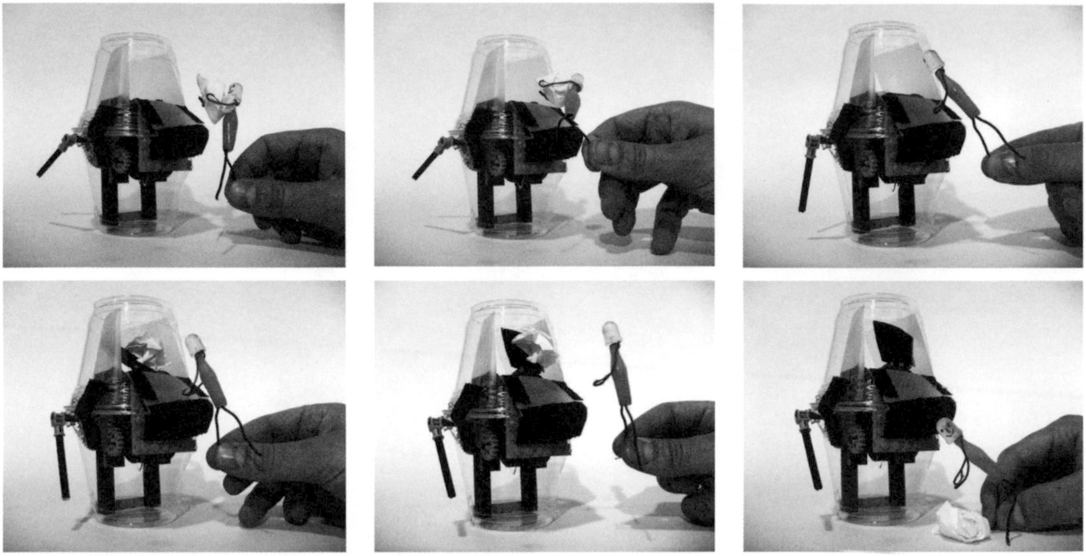

Lego prototype for designing Too Smart Trashcan, photography JooYoun Paek 2008.

with an object within it. Observing people using the informational signs, we noted that there was a tendency to first look at the sign from about three feet away before eventually moving closer and touching the signs.

Another interesting lesson from this research was discovering the rituals of sitting down in public spaces. While there are several benches scattered throughout Battery Park and the other sites we investigated, people had a tendency to sit on a wide variety of surfaces from steps to fencing. For most people observed, the most important factors in selecting a seat were the direction it faced and the distance it was from other people. When people did choose a bench to sit on, their period of occupation was both longer and more elaborate than when they had selected a non-traditional seating area. Their public ritual of sitting on a bench often involved placing bags around them, and stretching out, as if marking out this territory as theirs. During our research we did not discover any strangers sharing benches with one another. From this, we assume that benches are sites of temporary occupation that a person possesses as their own while seated upon them.

After finding a repeated behavior in public spaces, such as the rituals involving sitting on benches, our next step was to mimic the behavior as closely as possible. We would sit down on a bench and adjust our bodies into a position upon it. After finding this position, we would stand up and repeat, completing the act of sitting on the bench over and over in different positions. After going through this process, and attempting to dissect the behavior in our notes, we then deliberately engaged in the process incorrectly. This included sitting next to people on benches, or sitting as if on a bench in an area without one.

The entire activity of mimicry is a bit absurdist and performative, however, engaging in a process of both scientific observation and repeated performance did defamiliarize us from the ordinary activity. By the end of engaging in the mimicry process we were much better versed in the nuances involved in these everyday behaviors.

Prototyping

One of the chief difficulties in building advanced technological systems is that it is very difficult to change course midway. Aspects of the project, such as ordering custom mechanical parts for the bench, essentially lock you into a particular design choice. It is therefore imperative to first evaluate each of these choices as much as possible. One way of doing this is to build a series of prototypes that can help detect any design flaws with minimal investment in time and materials. For Too Smart City we had three stages of prototyping, first miniatures with Legos, a full scale version called Wizard of Oz and wood prototyping.

Legos are a wonderful tool for inexpensive tests of mechanical interactions because of the wide assortment of gears and levers that they provide. We began the process of building each of the furniture by building a miniature working

Lego prototype for designing Too Smart Sign, photography JooYoun Paek 2008.

Lego prototype version 1 for designing Too Smart Bench, photography JooYoun Paek 2008.

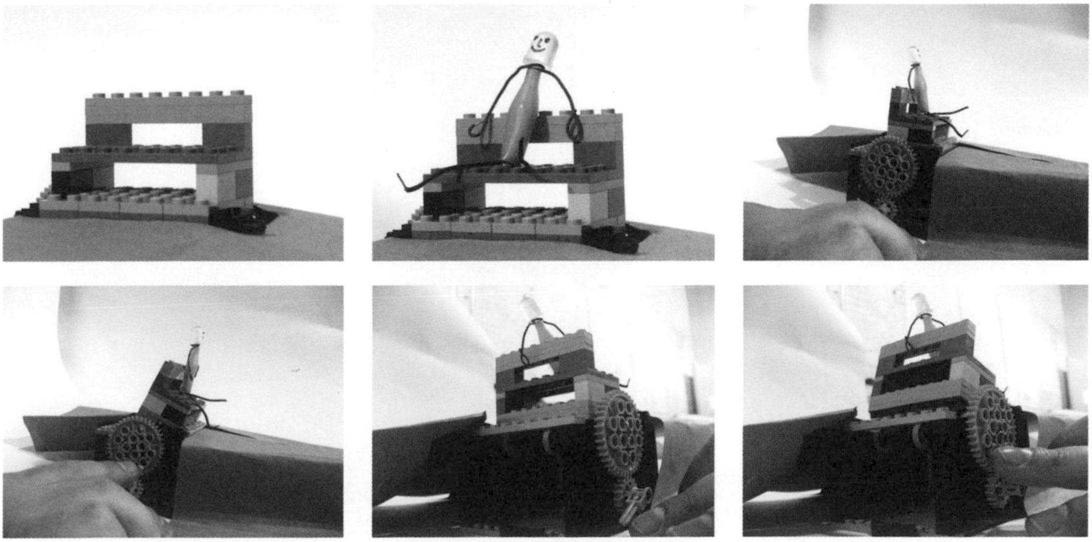

Lego prototype version 2 for designing Too Smart Bench, photography JooYoun Paek 2008.

Wizard of Oz testing props, photography by JooYoun Paek 2009.

Wizard of Oz testing, and Interview with subject after testing, photography JooYoun Paek 2009.

prototype out of Legos. The Legos were first assembled, and then reassembled with glue to keep the prototypes together across multiple uses. In these Lego prototype models, we concentrated on testing out some of our planned design concepts.

In the trashcan we modeled a rotating elevator platform, and then placed it inside a plastic cup with a lid, to simulate the general structure of the trash can. Outside of the cup was a lever system that allowed for us to operate the platform mechanism in a manner similar to how the motor would work. As we rotated the lever, the trash platform would first rise up and then throw the trash out. From this prototype, we learned that our plan for the platform would not work well, since trash was thrown directly back at the guest. Consequentially, we were able to redo our design concepts for the trash can without having spent any money on mechanical parts.

The Lego prototypes of the sign were helpful by simulating some of the weight and strain issues that the sign would have. Although the Lego plastic behaves differently than the aluminum of our final product, the prototype showed that having a long sign which was rotated and stopped at fast speeds would cause problems with torquing the motor. In addition, we were able to determine that our initial idea of having actual signs which mechanically flipped would require a structure that no longer looked similar enough to everyday signs. For the bench prototype, Legos allowed us to test out a few of our mechanical ideas. In addition to the actuators lifting the bench rear legs up, we had also conceived a bench with a conveyor belt system that would rotate until the person fell off. Mechanically, the conveyor belt system is superior, because it does not require shifting weight onto the two rear leg actuator motors. However, in our initial Lego tests, we found that there was a fundamental safety hazard with the conveyor belt because it pushed the guest off in a way that they could drop directly onto the floor. In contrast, by lifting its rear legs, the other bench design transformed into more of a slide. By adjusting the height the legs raised, we could determine the severity of the bench, assuring a level of safety.

Following our Lego designs, we built full scale prototypes of each of the furniture, for use in our "Wizard of Oz" prototypes. The Wizard of Oz methodology involves having the test operator perform a series of tasks behind the scenes, while guests go through and engage with the experience. It is a very helpful process to engage in because it enables testing core issues in the interaction and design upon a set of people, or testers, who can then give valuable feedback on their experiences.

Our prototypes were fashioned with available materials in order to quickly approximate the size and structure of the furniture. For instance, the trashcan was assembled out of cardboard boxes, with the operator hiding inside, acting as the mechanical system. When the tester threw trash into the receptacle, the operator would catch it, determine if it were metal or not, and then throw it back out. Wizard of Oz prototypes assist in adding a human element to the tests, and provide input from people who are not familiar with the designs. During our Wizard of Oz tests with the trashcan, it became apparent that people would throw away trash while walking past the Too Smart Trashcan, and have left the vicinity before the trash was thrown back out again. From our prototypes, we found that having a receptacle on the top, rather than on the side with a flap, allowed guests to look inside the mechanism, and upon seeing the laser, realize that the trashcan was not ordinary. In addition we determined that the response time between receiving trash and rejecting or accepting it was critical to promoting successful interactions. By watching videos made of the Wizard of Oz tests, we found that people tended to

Building wood prototype of Too Smart Bench, photography JooYoun Paek 2009.

Wood prototype user testing, photography JooYoun Paek 2009.

stand near the Too Smart Trashcan for an average of ten seconds.

As a final step in the bench prototype, we created a full scale mechanically operable version out of cheap lumber. This enabled us to ensure that our scale, movement, range and timing were all accurate before going to final fabrication. The joints between the moving parts were made to match those in the design documents to test out the mechanical system. In spending around one hundred dollars and three days labor, we were able to refine many of the design aspects. Unlike the other furniture, the bench has direct body contact with the audience, and therefore the feeling of sitting and moving with the human body was crucial to our design. We needed to find the sweet-spot between people remaining safe while feeling uncomfortable, as the bench lifted to its tilting angle. With the prototype we measured three major systems: the minimum tilt of the bench for users to feel uncomfortable and be forced to stand up; the speed with which to lift users, where they will be able to stand up without falling to their knees; and the minimum density of light sensors needed to have on the sitting board of the bench to accurately sense who and how many people are sitting on the bench. We performed all of these tests on a male and female of average weight and height to determine the final parameters in our design.

Designing and Building

The Too Smart City furniture needs to look like everyday public furniture in order for people to understand it as being part of a future everyday scenario. While, we are very familiar with the general designs of trashcans, benches, and signs, in order to get our look to be exactly right we needed to engage in more research.

The first problem was in selecting the style of public furniture we would copy in our designs. For instance there are wrought iron benches with curving armature, and there are contemporary benches of angular metallic structures. Fortunately, New York City has numerous examples of these designs across the five boroughs. For a period of three months, we photographed different furniture design forms, creating a repository of images that we then used to choose our designs. In addition, we measured public furniture, finding the exact dimensions of the designs being employed, so that we could match them more closely.

Eventually, our design selection was determined by which styles worked best across all three furniture. The polished aluminum that we chose was found in numerous public furniture examples, and could be adapted for use in the sign and the bench. Since only the internal mechanisms in the trashcan were moving, we decided to purchase the trashcan from a supplier, and then build a custom frame inside of it. The bench and the sign were both custom fabricated.

Conclusion

The public reception of Too Smart City was split between those who loved it, and those who did not understand that it was intended to fail. In our experiences, those who did not understand the art work were reacting to an article on the pieces, rather than having seen them in person. For instance the following commentary from AirPillo on Boing Boing:

I can just imagine my poor grandmother sitting on that bench and getting hurt when it tipped her off.

This is a rather common concern in the comment portion of the web, specifically grandmothers and the bench. It appears that we have hit upon a nerve specific to the elderly with respect to the mechanical bench. In the same article, Brainspore wrote:

I think it's safe to say that concept is intended as satire/social commentary. Why else would they call it "Too smart city?"

For us, this conversation reveals the success of the work. With all the effort to make the furniture seem like part of the everyday, to be nearly plausible, our greatest hope was that it would evoke a slight unease, as well as one of satirical humor. In this way, Too Smart City produces a conversation around the topic of the Sentient City, and encourages people to reflect upon it.

As interactive artists, the final consideration for Too Smart City is what its legacy will be. There are too many interactive works that end up relegated to storage once they are no longer perceived as novel or pertinent. For Too Smart City the issue for us is how the furniture can retain relevance after the Sentient City comes into existence. Obviously, within a few years the technology will be outdated, and the conversation may seem misguided or antiquated. For us, the meaning of Too Smart City is not its technology, but rather capturing a tension felt towards a possible future through comedy.

SITUATING KNOWLEDGE WORK IN CONTEMPORARY PUBLIC SPACES

BREAKOUT! ESCAPE FROM THE OFFICE

ANTHONY TOWNSEND, ANTONINA SIMETI, DANA SPIEGEL, LAURA FORLANO, AND TONY BACIGALUPO

ABSTRACT

Why do we work in office buildings?

This simple question was inspired by emerging trends that are creating new opportunities for situating knowledge work in the city. An increasingly mobile workforce is being attracted to a wide range of public and private work sites, from coffee shops to park benches. Ubiquitous wireless broadband, smart mobile devices, and location-aware social networks can activate almost any urban location as a potential worksite. New kinds of organizations, and a growing emphasis on cross-organizational and interdisciplinary collaborations is creating a greater demand for ad hoc face-to-face collaborative work in creative settings.

Recognizing these trends, and the deep historical relationship between work and the street that dominated before the 20th century's office building boom, the authors developed a set of tools and processes to support knowledge workgroups in urban public spaces. These tools and processes were intended to serve as an open source framework for exploring the new opportunities presented by mobile knowledge work for the utilization of urban space. The goal was to develop a global conversation around these practices that could be appropriated and adapted in urban contexts around the world, treating the diversity of sites, cultures and urban contexts as a rich set of experimental variables. These tools were developed and tested during September and October 2009 in New York City with a variety of locations, groups, and collaborative processes. Following their deployment in New York, these tools and processes were tested by Citilab, a project collaborator based in Cornellà, a city located on the outskirts of Barcelona, Spain.

The experience of Breakout! suggests future research questions and hints concerning how urban design, architecture, and organizational theory could respond to and drive re-integration of knowledge work and public space. We highlight three key areas of future exploration: catalyzing collaboration in the open source city, designing the mobile workplace, and rethinking the interface between office buildings and public space.

INSPIRATIONS

A broad range of emerging trends, projects and experiments inspired the conceptualization and development of Breakout! These inspirations fall into three main areas: historical relationships between work and urban public space, new social and connective technologies that support mobile work, and new forms of open source collaboration and organization of production.

Work and Urban Public Space

The form, function and identity of cities are to a great extent defined by the work we do in them. For most of the 5000-year history of cities, work was done in public spaces, where streets, markets and waterfronts served as platforms for commerce. A rich array of "third spaces" such as cafes, restaurants and taverns almost exclusively served the needs of businessmen. (Oldenburg, 1999) To take an example, New York City has a rich history of work in public spaces, and many important commercial institutions were developed in public settings. The founding agreement of the New York Stock Exchange, the 1792 Buttonwood Agreement, draws its name from the tree on Wall Street under which investors used to meet to buy and sell securities. After the Civil War, an informal community of curbstone brokers traded unlisted shares in the streets of the Financial District. This ragtag community of railroad and mining speculators would formally organize as the New York Curb Market in 1911, but didn't move indoors until 1921. In 1953, this once curbside financial market would become the American Stock Exchange. (Sobel, 2000)

Yet New York is unremarkable in this regard, as cities throughout the world display a close integration between the commercial spaces of the city and public spaces. But if we look at the patterns that dominated in the parts of cities built or rebuilt in the 20th century, there is a notable shift of work away from areas directly connected to public space, to be hidden away in high-rise office towers. How was this long established connection broken and what consequences have resulted?

There are many reasons for the migration of work from the streets and its segregation into private office buildings. Victorian mores certainly favored a separation of places for living and places for working. But it was the shift in urban economic activity from manufacturing and distribution to management and knowledge work that helped concentrate and separate work from the street. As factories were replaced by large bureaucratic organizations mainly engaged in paper-based information processing, office buildings provided the most efficient solution to bringing large numbers of people and large amounts of paper records together to support clerical work. One could say it was the combination of three transformative technologies — the steam engine, the elevator, and the telephone — that made it both necessary and possible to concentrate decision-making activities in high-rise office towers. (Gottman, 1983)

The office building served post-industrial cities well throughout the 20th century by providing a cost-effective, flexible gathering point for industrial companies. These organizations continued to organize their workforce around hierarchical management structures, capital-intensive and centralized business machines and infrastructure, and relatively non-transparent and impermeable interfaces to the outside world. Rather than seek a new architectural model, they simply upgraded business technologies — file rooms and mail rooms were replaced by mainframes and fiber optic connections. As architect Frank Duffy has written, the office building became a highly refined product — a formalized, standardized unit of urbanization with an entire industry and supply chain optimized to deliver it. (Duffy, 2008)

The future of the office building is only now beginning to be seriously rethought. Office utilization studies conducted by workplace design firm consultancy DEGW, a collaborator in Breakout!, indicate that even during business hours (just one-third of the hours in the 250 or so working days each year), modern offices are only occupied 30–40 percent of the time. This low utilization is the result of ongoing shifts in the way large organizations grow and change. The shortening of rapid product development cycles often means a greater reliance on collaborative partnerships and inter-firm networks over vertical integration, which leads to more time spent off-site.

Increased mobility through improvements in transportation and mobile communications allows numerous occupations, such as management, sales and consultancy, to work outside the office, and alongside clients, customers and subordinates at remote locations." Distributed infrastructure for computing and communications, printing and duplicating, and even prototyping and fabrication mean that central offices no longer have a monopoly on advanced business machines. Finally, the convergence of increasing mobility and social network technology has unleashed a surge of face-to-face business networking activity around conferences, workshops and retreats that serve to bring online communities of interest together for focused knowledge exchange.

The most progressive organizations are shifting their real estate strategies to reflect this emerging pattern, which combines high mobility, distributed communications, and decentralized meeting and collaboration spaces. Studies of corporations point towards increasingly mobile workforces, and the early adopters of distributed communications and collaboration tools like IBM and Cisco have driven their per worker space needs far below the 20th century norm. Many organizations, however, have not recognized this shift and continue to carry large inventories of office buildings, increasingly underutilized warehouses for workers. The problem is exacerbated by the fact that the commercial real estate development industry still has a supply chain optimized around the production of office buildings for tenants making long-term commitments to private space.

The final assault on the traditional office building is its growing ecological disadvantage. Commercial buildings represent a significant life-cycle contribution to global carbon emissions. In the United States alone, the building sector consumes 76 percent of all electricity and contributes 48 percent of national carbon dioxide emissions. (USGBC, 2010) Given that non-industrial, non-transportation commercial activity is 17 percent of all U.S. energy use, we can derive a ballpark estimate of 12 percent of electricity use and 8 percent of national carbon emissions shared between retailing and office buildings.

Worse yet, the typical corporate office environment is a study in creativity-killing design. Windowless conference rooms and desk cubicles focus teams on the task at hand, but eliminate the opportunity for outside information to orient or stimulate creative knowledge work. In contrast, the most attractive locations for mobile work — coffee shops, public plazas and parks, lobbies and restaurants — all offer dynamic venues for serendipitous visual stimulation and social interaction. Recent neurological studies have shown that human brains evolved through the stimulation of movement through forests, and the changing visual landscape of mobility is a powerful stimulant of awareness and problem solving. Breakout! was inspired by the possibility to not just change the economics of the workplace, but to exploit an opportunity to re-situate creative knowledge work in highly stimulating places that can support dynamic collaborations between organizations. The authors expected that public space would de-territorialize the workplace and make collaborations more open and equal than inside the offices of one of the collaborating organizations.

Taken together, this evidence suggests that new ways of structuring workspaces for collaborative knowledge work are likely to emerge. Since office towers are the defining and dominant element of city skylines today, the challenge of accommodating drastically different kinds of organizations in existing spaces will only grow. The shifts towards high mobility, and the increased need for collaboration between workers and organizations, creates an opportunity for public spaces and third spaces to become a part of the design solution for future workspace. This opportunity inspires us to look for precedents, both in the historical past of cities, and in the emerging logic of spatial organization in contemporary cities.

As we describe in detail later in this chapter, many new venues, tools and practices for collaborative work are being pioneered in urban public spaces by freelancers, small firms and mobile professionals who, for a variety of reasons, have liberated themselves from their offices. But for these pioneers, migration between ephemeral work sites in cafes, wireless parks and libraries has often devolved into an isolating experience. They may no longer need to co-locate in order to access shared tools and resources, but they are still drawn together by the need to socialize, interact, and collaborate. The rapid spread of coworking, where individual freelancers come together to fund shared workspaces for collaboration, is growing rapidly in response to the isolation of heavily virtualized forms of work.

But coworking, as successful as it has been, still clings to the 20th-century real estate-driven model of supporting work: that organizations need to be fixed, dedicated, and permanent to support creative and collaborative work. In that sense, coworking is partially locked into the real estate supply chain, and constrained by its economics. The idea for Breakout! was inspired by the challenges of implementing a coworking model pioneered in lower cost cities in the expensive risk-averse Manhattan commercial office market. By appropriating public space at no cost, the authors sought to offer a new economic model for coworking that removed real estate as a capital cost and operating expense. Through a timely coincidence, Breakout! was conceived just as New York City redesigned several large public streetscapes to reallocate land area devoted to vehicles and turn it into public plazas with tables, chairs and shade umbrellas, creating a network of experimental sites for the project.

A final aspect of the project's engagement with public space was the idea that workplace design has become locked into the assumption that all workplaces for skilled professionals must be situated within an office building. As we have indicated, many of these professionals have voted with their feet and chosen to work from client sites, vacation places, home, or Starbucks cafes. While collaborative knowledge work is now clearly the most productive activity of our society, the authors are not aware of any approach to creating collaborative workspaces that have truly re-conceived such spaces from an architectural tabula rasa. Breakout! frees us from the assumption that knowledge work requires single-purpose office buildings, and instead offers different assumptions to drive workplace design. For Breakout!, social networks and zones of mobility are the essential constraints, and the design challenge becomes one of tweaking public spaces that are highly accessible to mobile workers to accommodate multi-organizational teams on an ad hoc basis for spontaneous collaborative work. If the last decade was about rethinking the interior of the office workplace, over the next decade, we expect to see this critique expand to the shell and form of the office building and beyond.

Social Networks and Connectivity

Over the past decade, knowledge work has spilled out of office buildings and into public spaces supported by a host of communication technologies including hardware (laptop computers and mobile phones), software (social network applications such as Twitter and Foursquare and cloud computing applications) and infrastructure (Wi-Fi hotspots and cellular networks). Laptops have become more portable while mobile phones have become more powerful, expanding their functionality beyond voice communication and enabling text, e-mail, Web-browsing and a wide range of applications. These applications, which were ostensibly to be used for social rather than work purposes, have been essential for connecting dispersed groups of mobile professionals and freelancers. The widespread availability of free Wi-Fi networks in cafes, parks and public spaces, pioneered by community wireless organizations in the late 1990's, has become the essential infrastructure of the mobile office.

There are several competing visions about the ways in which these tools enable emergent forms of social organization. The techno-utopian vision of urban computing, which often dominates both academic and media accounts of technology, describes a world characterized by freedom from the office where creative and productive "anytime, anywhere" work is possible. Alternatively, the techno-dystopian vision laments the incessant ringing of the cell phone and checking of the Blackberry as signs of the encroachment of work into private time and public places including homes, vacations and other non-work settings. While Breakout! experiments with technologies that support knowledge work in public settings, by testing a wide range of socio-technical configurations, the project also attempts to strike a balance between the demands of work and life as they are embodied outside of the office.

In fact, in contrast to techno-utopian visions that advocate for complete freedom from place, academic research on social media and mobile work suggests that the availability of new tools for connectivity has made the role of place even more relevant. While early on the social, political and economic transformations enabled by the Internet shifted the focus of discussion to online, virtual and digital associations, scholars examining mobile technologies, Wi-Fi networks and location-based social media have documented the ways in which these tools have re-configured our attachment (Hennion, 2004) to people, technologies and places.

This may seem to be an obvious point in a book about situated technologies and sentient cities. Yet, it is one that is often glossed over in compelling discussions about spaces of flows (Castells, 1996) and frictionless cities. Fundamentally, people are sticky. Not in the economic sense of the word, which suggests a lack of change, but rather in the associative meaning. We move around together, attracted by the presence of others. The image of the lone mobile worker, whether on the beach or in an airport, is incorrect and misleading. While connectivity is one driver of foot traffic, the availability of good coffee, shade or other people is just as likely to attract.

For example, Humphreys suggests that Dodgeball (a text-based precursor to Foursquare) has enabled a "social molecularization" as users move about the city in small ad-hoc groups (2008). Similarly, rather than embracing the freedom to work anywhere, mobile professionals use mobile social networks to recreate routines, seek out interactions, and appropriate spaces in ways that allow them to be productive and creative. (Forlano, 2009) While mobile work practices have become more pervasive and evenly distributed, many of these habits were initiated by members of geek-publics (Powell, 2009) that lead the way in both innovating and using (Von Hippel, 1978, 2005) social and technological infrastructures that support knowledge work.

Based on this understanding of the relationship between connectivity and emergent organizational forms, Breakout! designed a technological intervention that built on and enhanced the existing communication tools for mobile professionals. The technological intervention combined a portable Wi-Fi hotspot, a communication platform and a mobile office kit. This socio-technical arrangement allowed for a range of combinations, which were modified throughout the project.

For example, some events used the Wi-Fi heavily while others did not use it at all; alternatively, some events relied on the publicly-available Wi-Fi from a park or store nearby. Similarly, the communication platform or Breakout! Operating System, which allowed Twitter-feeds to be streamed to a single site during an event, required a great deal of engagement and, thus, was often used more actively by some participants or during certain events. Finally, the mobile office kit consisted of an assortment of symbolic, designed objects and technical artifacts that are commonly used to support collaboration (sticky notes, white boards), simulate privacy (noise cancellation headphones) and maintain connectivity (Skype headset).

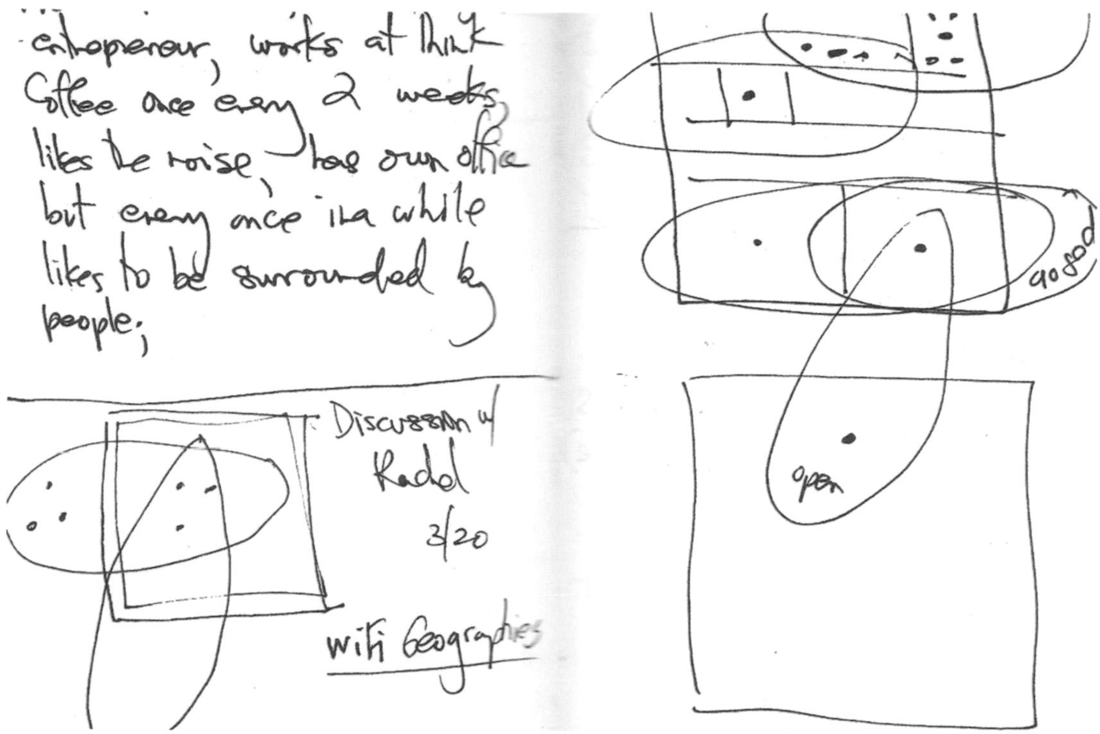

entrepreneur, works at Think Coffee once every 2 weeks, likes the noise, has own office but every once in a while likes to be surrounded by people;
Discussion w/ Rachel 3/20
WiFi Geographies
open

Mapping the geographies of WiFi networks in Laura Forlano's sketchbook.

Jessica, a self-employed graphic illustrator and entrepreneur, arrives at Think Coffee, a coffee shop in the West Village near New York University, at 10am on a rainy Friday in mid-March. She chooses a small round table along the back wall close to the electricity outlets where she will spend the majority of the day. Jessica has an office for her small business but likes to work at the café about once every two weeks. She likes to be surrounded by people and the lively din of biology study groups, business meetings and intimate conversations interspersed with the shrieking of the espresso press and muffled by a blanket of alternative rock, which plays on the overhead speakers. "I like the noise, if that makes any sense," she comments.

10:00 AM

It is stifling hot and steamy in the café, the temperature elevated by the density of people and laptops along with the bustle of activity in the open kitchen. Jessica takes off her navy pea coat and a black wool sweater in order to feel more comfortable. She goes to pick up the small take-out coffee that she had ordered on the way in adding a little whole milk and two packets of white sugar. Once back at the table, she takes out her laptop, a portable mouse, her cell phone, an electronic calculator and a bottle of water, which she spreads out in front of her. She logs onto Think's free WiFi network, updates her Facebook and Twitter social networking status updates to read "Working at Think Coffee today," checks e-mail and reads a few posts from Core 77, a popular industrial design blog.

10:43 AM

Jessica takes the small clamshell phone and heads towards the front entrance of the coffee shop. It is too loud inside and she needs some privacy in order to follow up with one of her clients about a Web site that she is designing. While gazing at the rainy streets, wet yellow taxis carve creative pathways through her mind. After well over an hour in the small square vestibule — which doubles as a bright red English telephone booth — Jessica grabs a chicken caesar salad and heads back to her table. It is noon and the café has gotten crowded.

12:08 PM

Around 1pm, Jessica's friend Andie stops by to check out her latest designs and offer some feedback. "I read on Twitter that you were here. How is your work going today?" Andie asks. Jessica and Andie met at Think Coffee a few months ago while waiting in line for a table and quickly became colleagues and friends. They like working at Think because they feel that they are better able to focus on their work when they are around other people who are working. While Jessica likes the noise, Andie uses noise cancellation headphones in order to concentrate on her writing.

1:18 PM

A thirty-something man sitting next to Jessica looks up from his typing. "Does anyone know a good database programmer?" he asks. "One of my clients just sent me a posting about a new job opportunity." Jessica gives him her business card saying that her husband is a database consultant. She keeps a stack of business cards in her purse and gives them out frequently at the café. She's even gotten a client or two this way.

2:36 PM

Jessica is getting tired and uncomfortable in the wooden chair where she has been perched for the past several hours. She decides it's time for a short walk to get her ideas flowing again. She decides to stop by New Work City, a coworking community on Varick and Houston, where her friend is a member. "Before you go, can you watch my laptop while I go to the bathroom?" Andie inquires. She's forgotten her security laptop lock today and doesn't want to lose her coveted seat.

3:21 PM

A day in the life of a mobile professional at Think Coffee.

BREAKOUT
ESCAPE FROM THE OFFICE
Login or Register
Barnes & Noble - Union Square
About This Location
Session Archive
Anthony T.: eating kimchee jigae, pretending im in Seoul already, remembering how to speak restaurant Korean.
20 hours, 51 minutes ago on Twitter
Mark B.: I just unlocked the "Local" badge on @foursquare! http://bit.ly/qPu3X
20 hours, 52 minutes ago on Twitter
Rob K.: check out the Breakout OS beta session http://bit.ly/11tzW0 (i'm checked in remotely) @breakoutnow
20 hours, 53 minutes ago on Twitter
Anthony T.: Breakout beta session report - today we had the third "beta" Breakout session, where our team and friends are... http://tumblr.com/x0q2vmou2
21 hours, 29 minutes ago on Twitter
Mark B.: PAHAHAH! RT @kibbe: So glad I'm a grown up. This is how grown ups put appointments in their iCals: http://twitpic.com/fl4bk
22 hours, 8 minutes ago on Twitter
Anthony T.: the Cradlepoint battery is dying, survived for 2.5 hours with 4-5 people wailing on it
22 hours, 16 minutes ago on Twitter
Anthony T.: whiping out the netbook for @epreposi cause her battery died. you know your a geek when you keep a spare computer in your pocket
22 hours, 18 minutes ago on Twitter
Antonina S.: trying to make this work...
22 hours, 20 minutes ago on Twitter
Anthony T.: having a show and tell of Breakout tools - Mimo USB Mini Monitor, Cradlepoint battery-powered 3G-Wifi router @breakoutnow
22 hours, 21 minutes ago on Twitter
Elysse P.: Breaking out @ Barnes and Noble in Union Square
22 hours, 21 minutes ago on Twitter
Laura F.: prepping for New Media and Global Affairs course
22 hours, 23 minutes ago on Twitter
Mark B.: At @breakoutnow in Union Sq - Photo: http://bkite.com/0bgic
22 hours, 24 minutes ago on Twitter
Laura F.: checking in w/ @anthonymobile @danaspiegel @elysseprepos
22 hours, 29 minutes ago on Twitter
Dana S.: @breakoutnow (@ Barnes & Noble - Union Square in NYC w/ @markb) http://bit.ly/GdlmO
22 hours, 48 minutes ago on Twitter
Anthony T.: the BreakoutOS is alive and we're checking in
22 hours, 52 minutes ago on Twitter
Host a Breakout Session
About This Location
33 East 17th St, 3rd Floor
New York, NY 10003
Here Now
Anthony T.
Dana S.
Laura F.
Elysse P.
Mark B.
Antonina S.
Rob K.
1 Past Sessions Here
Breakout Beta Session from 4 p.m.-7 p.m. on Fri, Aug 28
View More »
SPONSORS
THE ARCHITECTURAL LEAGUE NY
nycwireless
Follow Sponsor Contact Archives
Creative Commons Attribution—Noncommercial—Share Alike 3.0 | Terms of Service | Privacy Policy

Status updates from Twitter-feed on Breakout! Operating System from a beta session.

By using a portable Wi-Fi hotspot rather than relying on the existing communication infrastructure available in public spaces, it was possible to expand the network of spaces that were occupied during the events. It is no surprise that Wi-Fi played a central role in the orchestration of the events as the team included a deep knowledge of wireless technologies gleaned from nearly a decade of work with NYCwireless, the non-profit organization responsible for many of New York City's free, public park hotspots. Again, rather than the goal of ubiquitous connectivity, Breakout! explored the opportunities and constraints of bringing Wi-Fi into uncharted spaces.

As cities are increasingly populated with the devices, data and systems that embed knowledge work in public spaces, it is necessary to question the motivations and perceived benefits of these technologies. To paraphrase an academic colleague, it is necessary to be both wildly enthusiastic and, at the same time, appropriately frightened by the capabilities and consequences of the choices that our society is making when it comes to urban informatics. While Breakout! sought to enhance the quality of knowledge work by using technology to assist in the formation of new associations, ideas and networks, it is important not to overlook the potential dark side of unleashing a flood of mobile workers into public spaces.

Most cities have not done enough to provide the critical infrastructures of the 21st century. If nothing else, Breakout! cracked through the glazed windows of the post-industrial age in ways that only hint at the possible future of work.

New Forms of Collaboration

There has been much optimism about the success of the Free Libre / open source software movement and the benefits of peer production over the past two decades. This emergent mode of production, collaboration and innovation has expanded into fields far beyond software and computer technology — the building blocks of so-called digital, virtual and online spaces — and into industries including biotechnology and pharmaceuticals. It has also inspired the development of the do-it-yourself (DIY) movement and of business models for public services, such as car-sharing, clothes-renting and book-swapping. These services employ digital technology to facilitate the sharing and trading of physical artifacts and commercial products. All of these practices require a movement away from proprietary forms of intellectual property and towards sharing and collaboration as an important mode of engagement with ideas and people. How might the application of peer production to emergent, mobile work practices facilitate new kinds of collaboration and innovation? In order to answer these questions, it is necessary to better understand emergent organizational models such as co-housing and coworking.

Apart from changes in the physical arrangement of cities and offices, new forms of production and social organization have emerged. In particular, there is great interest in the potential of peer production and the sharing economy. (Benkler, 2006) Just as we have witnessed the success of collaborative projects such as Linux and Wikipedia, there is significant momentum around initiatives such as the University of the People and projects such as the Public School for Architecture, which aim to expand the boundaries of traditional knowledge based in institutions such as universities by introducing peer-produced courses supported by networks of volunteers. These efforts also have their potential constraints and drawbacks as some fear the devaluing and deskilling of professions that have traditionally required specialized expertise. At the same time, informal learning ecologies and communities of practice are believed to be important in education and innovation.

Lead users (Von Hippel, 1978, 2005) and user entrepreneurs (Tripsas & Shah, 2007) are believed to be fundamental to the creation of new products, markets and services in a wide range of areas from machinery to technology and consumer goods. Mobile workers are a kind of lead user in that they are constantly testing out new ways of working with the support of technology in a wide range of non-traditional office settings. For example, earlier research by Forlano (2008) described the ways in which mobile workers appropriated semi-public spaces such as cafes as temporary incubators to get feedback on their work from passersby. Organizing diversity (Girard & Stark, 2002) and weak ties (Uzzi & Dunlap, 2005) are also believed to enhance the innovation potential of traditional firms. Finally, informal interactions (or 'water-cooler' conversations) contribute to collaboration and innovation in traditional office settings through the building of trust and social support. However,

An example of virtual communication: Breakout! team meeting with Anthony Townsend and Dennis Crowley pictured and Georgia Borden and Sean Savage on screen using DEGW's videoconferencing system.

research has shown that as the use of e-mail increases, informal communication decreases (Sarbaugh-Thompson & Feldman, 1998).

Unlike much of the literature about virtual forms of organizing, which focuses on collaboration in online settings, Breakout! created interventions into face-to-face settings, which were augmented, extended and enhanced by the use of communication technologies. Rather than screen-to-screen interactions between people separated by great distances, Breakout! privileged the dense informal interactions that occur when people are within close physical proximity. This juxtaposition of people, technologies and places suggests a hybrid social format in which the balance between face-to-face communication and screen-to-face communication is reconfigured for the purposes of greater awareness of ongoing activities.

How might these interactions trigger new face-to-face connections and associations when broadcast in real-time as status updates? How might they provide a record of ongoing conversations for future reference? What is the equivalent of the water cooler in mobile work settings and public spaces? With these questions in mind, Breakout! set out to experiment with emergent forms of organizing such as mobile work and coworking (working side by side but not together on a shared task), as well as meeting formats, methods and exercises that have become popular in technology world, such as BarCamps, pair-programming, speed-dating, lightning talks and flash mobs (Rheingold, 2003).

BarCamps, or un-conferences, are very loosely structured, ad hoc gatherings of

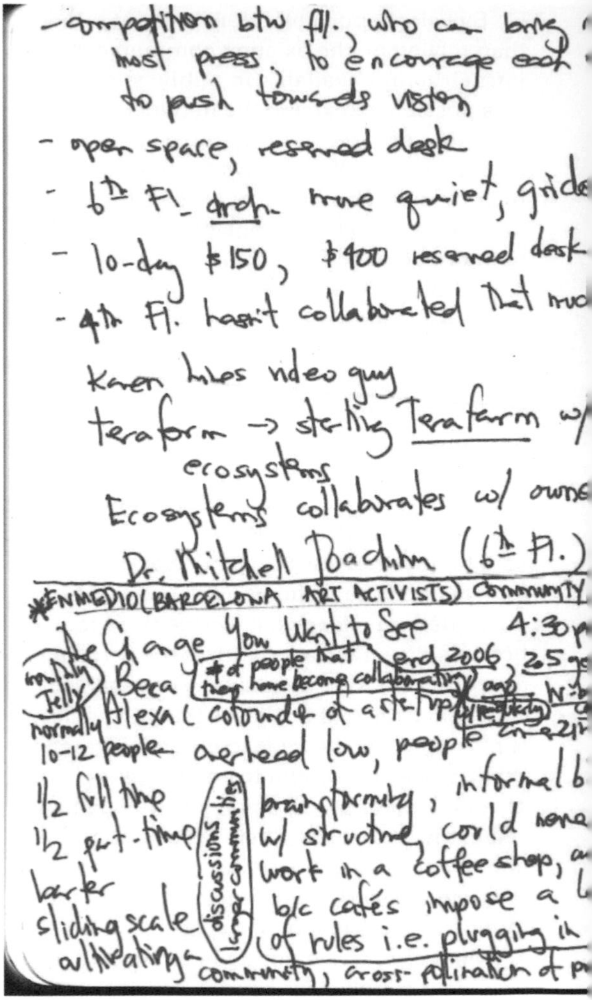

Field notes from tour of coworking spaces in Brooklyn organized by Todd Sundsted.

people who want to share information, socialize, teach and learn in an open environment. The BarCamp agenda is set by the participants at the beginning of the event and modified as necessary throughout the day. In this format, all participants are expected to actively contribute and lurking and self-promotion are frowned upon.

Pair-programming is a method used in software development in which teams of developers work closely together. In this mode, one person takes the role of "driver," focusing on the tactical minutiae of the work, and the other takes the role of "navigator," by observing, catching errors, making suggestions and thinking about the strategic direction of the work. Just as outsiders to a field have often contributed some of the most important innovations, pair-programming takes advantage of the fact that navigators often have the

Brainstorming as part of the "Future of Mobile Work" session.

ability to "see" important aspects of a project that drivers have missed since they are deeply immersed in the details. While this format has been useful in the software world, it also offers a possible model of collaboration for other fields including designers, writers, chefs and accountants as well.

Speed-dating and lightning talks are, perhaps, more familiar meeting formats. In speed dating, a circle of chairs and tables is set up and participants rotate through the full circle giving two-minute introductions to each person in the group. Speed-dating facilitates one-on-one networking and is a good way to ensure that everyone in the group gets to chat briefly with everyone else. For focused theme-driven discussions, lightning talks challenge participants to present a short talk accompanied by a slideshow with a predetermined time limit. For example, Ignite, an O'Reilly geek event with the motto "Enlighten us, but make it quick," allows presenters 20 slides that auto-advance every 15 seconds for a total of five minutes.

The majority of Breakout! events fell into the coworking, BarCamp and flash mob categories, in part due to the interests of team members and participants. The coworking events invited participants to work individually alongside new people and in experimental settings. The design charettes and mobile office design events, which are described later in this chapter, followed the BarCamp format most closely. Finally, applying ideas about peer production to social science methods, the Flash Mob Ethnography sessions mobilized teams of designers, technologists and researchers to gather data about the economic recession in the Union Square area. In about an hour, the teams collected over 260 rich images, which were collaboratively coded and described in subsequent sessions.

By experimenting with these new modes of collaboration supported by communication technologies and across a range of settings, Breakout! gained insight about emergent work styles as well as the opportunities and constraints of these practices. These insights will be described in greater detail below in the discussion of "Collaboration in the Open Source City."

INSIGHTS

Based on these inspirations, the authors developed a program of interventions in public spaces which were conducted in New York City in September and October 2009. This section describes these experiences in greater detail and the key insights about the emerging relationship between work and the city.

Collaboration in The Open Source City
Drawing on Duffy's earlier research on the changing nature of work and cities (2008), Breakout! explored the boundaries between offices and cities, work and play, and co-presence and absence. Yet, while Duffy's argument focuses on the reorganization of cities to support environmental sustainability, and the multifarious reasons why this has not been possible to date, Breakout! is concerned with the activation of ideas, creativity and spontaneity in knowledge work. Inspired by the grassroots organization of a plethora of coworking spaces throughout New York and the world (Forlano, 2009b), Breakout! asks what it might mean to live and work in an "open source city."

Just as Internet enthusiasts have lauded the great accomplishments of virtual communities around projects such as Wikipedia and Linux, the oft-cited poster children for online collaboration, a wide range of social media are now available for hyperlocal-problem solving, information and news. These include FixMyStreet (a site that allows citizens to report local problems), Outside.in (a hyperlocal blog aggregator) and Groundcrew (a mobile platform for local activism and engagement). As networked digital technologies such as wireless sensors, augmented reality and interactive urban screens play a greater role in our physical environments, we will begin to see the ways in which collaborative modes of production in online communities can be translated to what we have previously thought of as offline settings.

Along these lines, Breakout! demonstrated the potential for real-time problem-solving by ad hoc groups in dense, urban settings. It highlighted the ways in which collaborative activities might be designed to activate like-minded groups. It recognized that complementary skills and heterogeneous networks are necessary for collaboration. It leveraged social media to bring people together in face-to-face settings and document their activities.

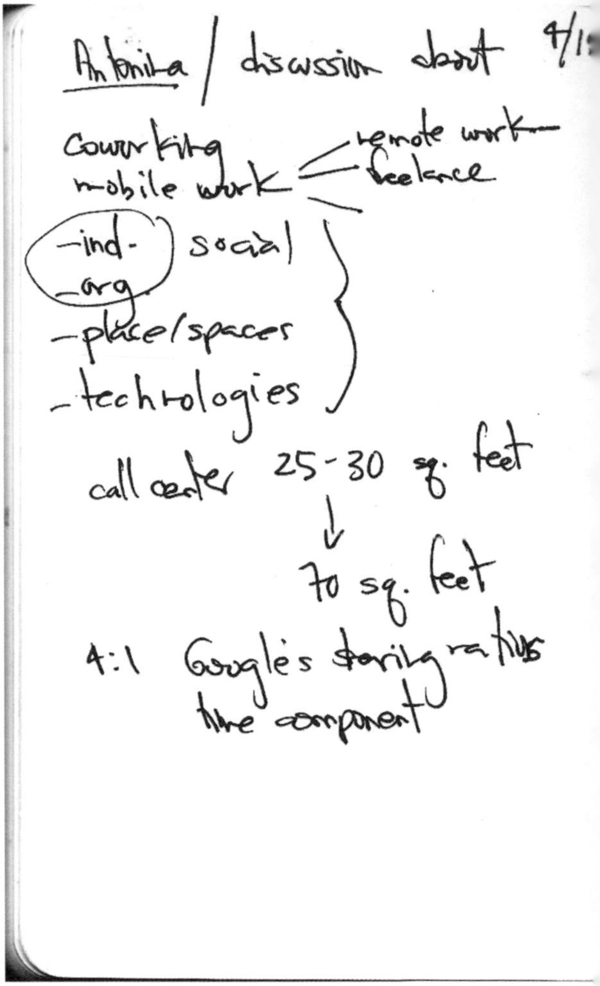

Field notes and diagram on key concepts from Laura Forlano's sketchbook

Yet, at the same time, the project uncovered a number of challenges in realizing its vision for social change: first, the majority of knowledge workers are not mobile; second, the use of social media is not evenly distributed; third, for the most part, people have not been trained to collaborate; and, fourth, we have not properly evaluated the potential societal risks (not to mention the damage done to minds and bodies) to the overflow of work activities into public settings.

First, the majority of knowledge workers are still bound by the schedules and seating-plans of the post-industrial age. While mobile workers are an emergent category, it is still difficult to articulate who they are, who is included and who is excluded. Are they freelancers or executives, entrepreneurs or academics? Or, are they anyone who has control over when and where they work? Or, anyone that can escape their

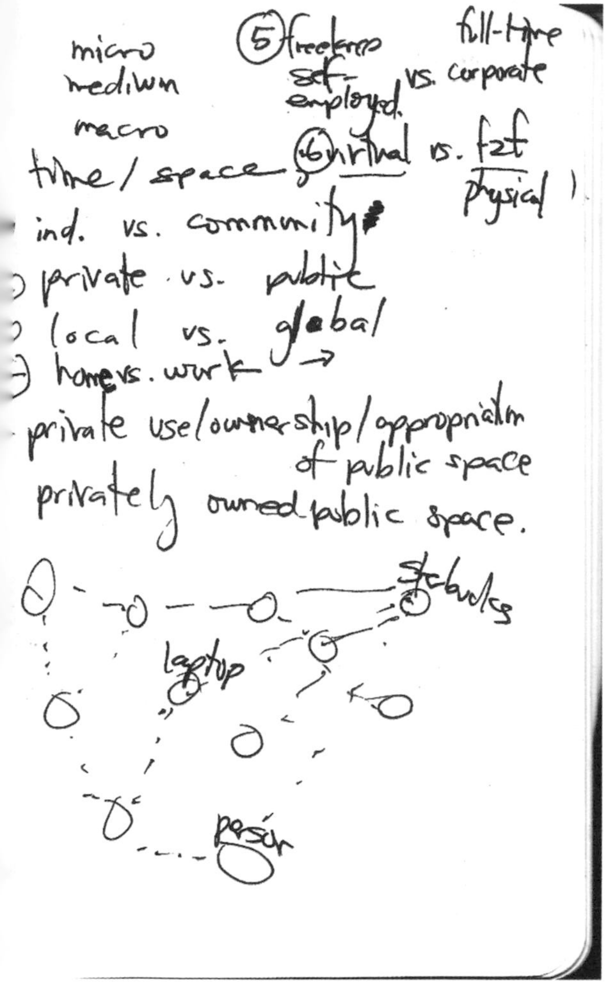

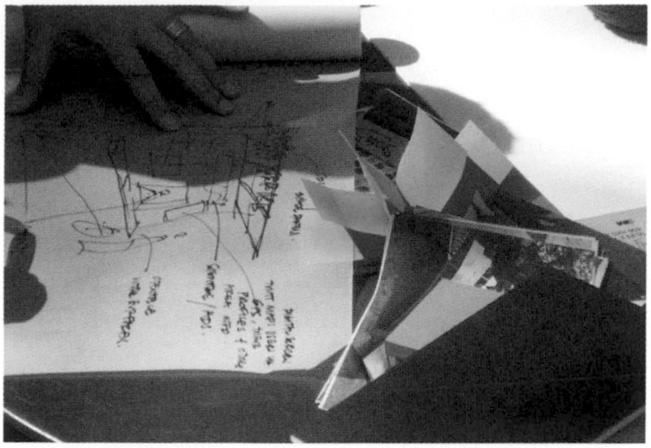
A designer sketching future work scenarios as part of the "Mobile Office Design" session.

usual routine for a few hours? In what industries do they work? Technology, graphic design, journalism, consulting? Breakout! struggled considerably with questions about the definition of this group. In the end, if nothing else, we can agree that mobile workers are a new constituency, held together by the common practices of having flexibility over when and where they work. But, how can we mobilize this new constituency? What values do they hold in common?

Next, while social media such as Facebook and Twitter have been heavily embraced by some groups, they are not used at all by others. In particular, while Facebook has become relatively mainstream among certain demographics, Twitter is limited to a more selective group. Status updates have been found to have great utility for sharing professional expertise, commenting on news and networking for certain individuals, industries and professions. Thus, when you isolate both mobile workers and Twitter users in a given geographic area, you reduce the number of potential participants considerably. For example, signing up on the Breakout! platform with a Twitter account was a barrier to participation. This fact was perhaps not adequately considered in the planning stages of the project.

Next, the issue of collaboration. While online collaboration and crowdsourcing have made headlines for their potential to generate value for society as well as for private companies, most collaborative projects, including Wikipedia and Linux, require multiple levels of engagement, trust and commitment. In some professions, such as large-scale scientific research or architecture, collaboration is considered to be essential whereas in others it is relatively unexplored. In either case, our education system continues to train students to focus on their individual projects rather than how to participate in multi-disciplinary, collaborative teams. In the case of the Breakout! project, guided collaborative activities were more generative in ideas and outputs versus events in which collaboration was expected to emerge spontaneously.

Finally, Breakout! embraced a vision of society in which mobile workers connect and disengage free from the constraints of the post-industrial office. As we have discussed earlier, rather than exemplifying a completely new practice, in some ways, mobile work has returned us to an earlier era in which much of working life in some professions, took place in the streets. At the same time, this re-emergence of working in public spaces has been augmented and reconfigured due to the availability of communication technology. As a result, we must carefully consider the

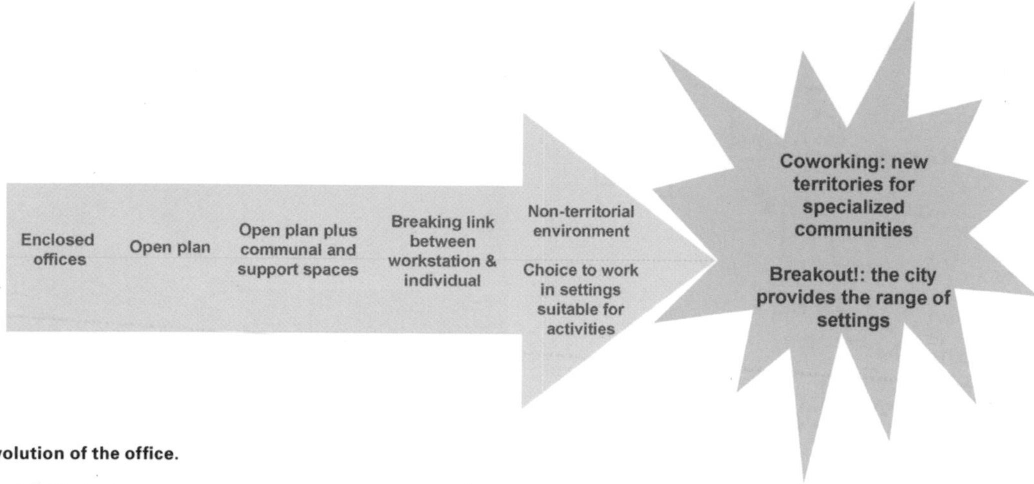

Evolution of the office.

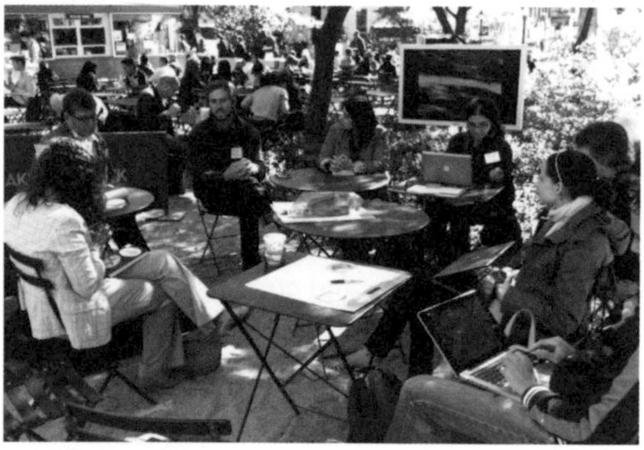

A Breakout! session explores design concepts for public workplaces in situ.

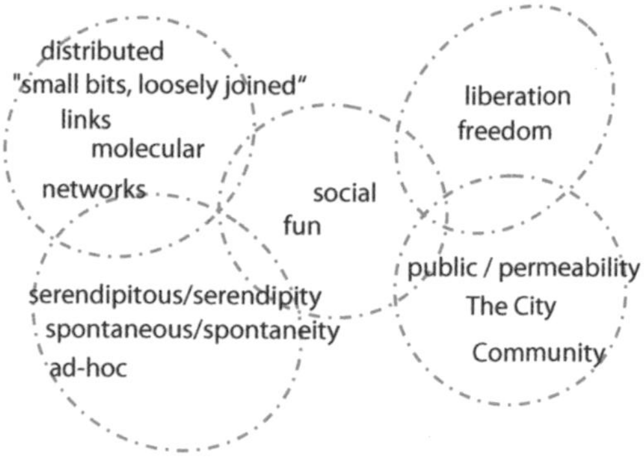

Team vision for the Breakout! festival.

benefits and risks of unbridled work creeping into times and places previously reserved for relaxation, unwinding and play.

From one perspective, the "cat is already out of the bag" so to speak. With the widespread use of the mobile phone and Wi-Fi, social norms have already shifted considerably to allow and even encourage the presence of work in all manner of places. At the same time, we are at a critical moment for designing the kinds of activities that we would like to encourage and, even, dissuading other activities. Rather than merely greasing the machines of capitalist production or focusing on the "work" that we do to pay the bills, Breakout! seeks to encourage the kind of work, whether it be paid or voluntaristic, that we do to build a better society. It is this kind of work that will bring the vision of the open source city to fruition.

Designing the Mobile Workplace

In planning to situate work in public spaces, Breakout! drew upon corporate workplace design, a field rich with urban metaphors and concepts. As a way to inspire serendipitous interactions, idea sharing and innovation in the workplace, space planners have appropriated the vocabulary of urban design to define office environments that are easily legible by users and cue those familiar activities that occur spontaneously in the city (e.g. plazas or town squares for meetings and socializing). In contrast, Breakout! returns work to the streets of the city, mapping the office to urban places and applying workplace concepts to public spaces to encourage community and interaction (e.g. a park as a water cooler for informal exchanges of ideas and information).

With Breakout!, the city is the office. Where are the best urban spaces for work — for meeting, concentrating or presenting? In order to answer these questions, the Breakout! team engaged in a collaborative design process that included building an understanding of the evolution of office design, conducting user research, visiting coworking sites, and

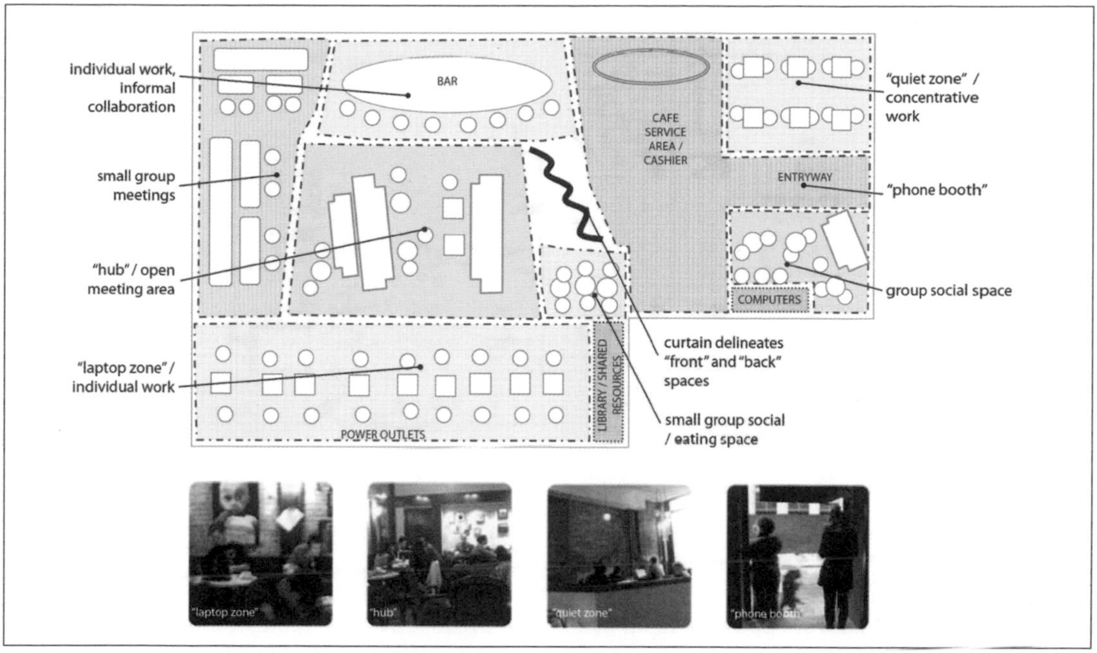

Mapping work activities at Think Coffee to illustrate how coworkers and students reappropriate urban spaces.

participating in collaborative design charrettes to create designs for new mobile work environments for the city. Breakout! was a collaborative process itself as much as it was a set of software, tools and events to exhibit mobile work in the city. The project put into practice the very same collaborative ways of working that it sought to explore and improve.

Over time, the corporate workplace has progressed from partitioned and assigned spaces to environments that are more open, distributed and lacking individual "ownership." Where do coworking and Breakout! fit into this evolution? The infinite choices of where to work provided by mobile technology make the network of spaces and amenities around the place people choose to work more important. While technology has "untethered" workers from space, coworking reemphasizes the value of location and puts like-minded communities back in specific workplaces. Coworking localizes the benefits of the agglomeration of specialized work — knowledge spillovers, proximity to collaborators, and access to support services — by clustering specialized communities in more intimate and relevant environments.

Breakout! goes further by placing knowledge workers into a significant and valuable context rather than creating the space — taking advantage of the community, amenities and infrastructure already provided by neighborhoods in the city to position a workspace and collaborative activities that benefit from the immediate environment. For example, Breakout! invited designers to Madison Square Park (a public space located in a neighborhood with a concentration of design firms), to contemplate and brainstorm ideas for portable office structures and furniture. The environment was the source of inspiration, allowing the designers to collaborate in the very place where the design ideas might be implemented, and simultaneously challenged the assumption that all design work requires a specific set of permanent, immobile tools and infrastructure.

As discussed earlier, Breakout! assumes that coworking is a model that defines the future direction of work, and therefore uses coworking spaces, activities, and technology as a starting point for inspiring Breakout! designs. The team conducted research on the network of coworking sites in New York City to learn about the experience of knowledge workers in existing spaces, collaboration activities practiced there, and user preferences for furniture and infrastructure. The research informed concepts for the mobile office, but also informed the design of Breakout! software, the location of Breakout! activities, and best practices for Breakout! sessions throughout the festival.

The team surveyed a range of coworking/mobile work environments across New York City through space observations and interviews with space operators and users. The spaces

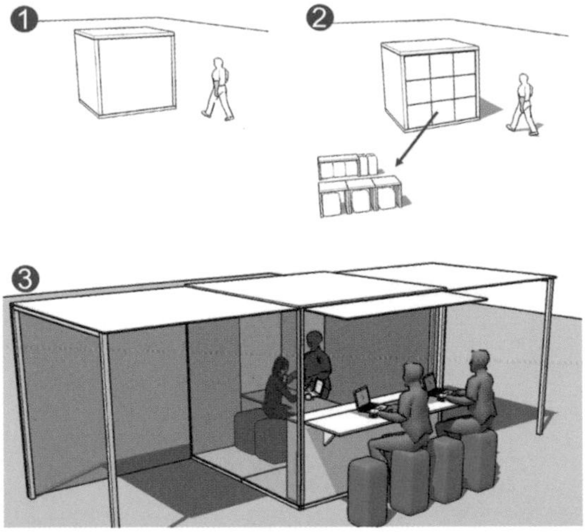

The Cube: Turns a plaza into a complete and modular work setting.

Circles & Squares: Adaptable furniture for a range of work activities.

visited ranged in formality — from a café with a set of regular mobile work patrons, to an incubator space for high-tech start-ups. The community of mobile workers at each site visited was distinct in the formality of its membership (public and open versus structured membership), industry focus (software development, journalism, or design), and stage in the business development process (conceptual versus incorporated and seeking capital). Despite this variety, a set of common space types and ways of collaborating emerged that became the framework for the Breakout! mobile office design considerations. The Think Coffee café environment — the least formal and most public workspace visited — illustrates the way in which mobile workers can easily appropriate space for diverse independent and collaborative work activities.

The range of mobile worker activities defined the zones. A set of common, simple and moveable furniture pieces (café tables, chairs and sofas), together with more permanent design features (bar, curtain wall, elevated seating area), provide different settings to meet the degrees of concentration, collaboration and privacy needed.

To develop a vision for the physical context of Breakout!, the project team explored and defined the goals and parameters of the festival through a series of design charrettes. The following set of adjectives used to define Breakout! drove the design of the mobile infrastructure, collaboration software platform, and events.

The festival vision, together with findings from coworking site visits, served as the basis for the design principles for Breakout! architecture:

— *Support a variety of work activities (individual, collaborative, sharing, and social)*
— *Enhance and activate public space*
— *Be replicable and affordable*
— *Be iconic*
— *Build upon existing site amenities and infrastructure*
— *Provide shade, privacy (as needed), and comfort*
— *Be assembled daily by two–three people*
— *Support collaborative sessions of up to 20 people*
— *Function without electrical power*

From these principles and constraints, four concepts were developed to provide a range of solutions for situated knowledge work in public spaces. These designs featured varying scales and degrees of portability, for defining or circumscribing workspaces, and supporting a range of work activities.

The Cube

The Cube is a heavy-weight design intervention for a "blank slate" site with no existing infrastructure. The architecture explodes the standard 8′ x 8′ office cubicle to create an expanding and unfolding container that defines the edges of the workspace, and provides shade, work surfaces, seating, and projection / whiteboard walls. The Cube provides a range of work settings and configurations that are controlled by the user, and can support meetings,

Cobrella: Define a site with shade and screens.

The Mobile Office Kit.

individual work and social events. The Cube is a self-contained package including the structure and all furniture. The structure is meant for one-time installation at the start of the festival with minimal reconfiguration, and can provide opportunities for branding and public art on its exterior when it is in a closed position.

Circles and Squares

This medium-weight design intervention creates a "café-like" environment. Circles and Squares is a set of flexible, branded furniture that can be transported between project sites. The furniture module is an adjustable "square" — one surface that slides to various heights in order to support different activities on-demand. The furniture can be dropped into "blank slate" sites or sites with some existing street furniture. Circles and Squares provides shade, surface, and seating, is quick to setup and easy to reconfigure by users.

Cobrella

Cobrella is a lightweight installation designed to outline a Breakout! work environment. The structure integrates with existing street furniture, carving out the desired workspace in the existing area. Cobrella can quickly be set up to fit a variety of site shapes / dimensions and easily reconfigures to define spaces varying in size and openness. The structure provides shade and adaptable walls for privacy and projection when needed.

The Mobile Office Kit

The Breakout! team curated a Mobile Office Kit for use during the festival. The Kit was the most lightweight and portable intervention proposed, and an "anti-architecture" solution for Breakout! that optimized mobility and portability. The Kit includes a set of traditional tools to support collaboration (e.g. Post-its), tools to delineate space (e.g. street chalk), and artifacts to help create visual cues for the nature of activities taking place in the space (e.g. red "concentration" cap).

The Kit solution puts greater emphasis on context — the selection of the location, participants and the type of collaborative activity planned for the Breakout! session. The City's urban places typically provide much of the infrastructure needed to do work — writing/ laptop surfaces, seating, vertical presentation / projection surfaces, and wireless access (provided by the park or via portable router) — and Breakout! relies on the City to supply the resources necessary to support remote workers.

Design thinking continued throughout the festival, in the form of collaborative "Breakout! Sessions" to explore mobile work practices and the appropriate urban environments for it. Part of a larger program of events, these sessions provided focal points for participants to explore and reflect on the physical design considerations of work in public spaces.

The Future of Mobile Work

60 Wall Street public atrium, New York City
September 24, 2009
18 attendees took part in this Breakout!

session, which brought together mobile workers / co-workers and corporate workers for a conversation about the future of mobile work. Participants included workplace strategists from innovative corporate organizations (such as Accenture, American Express and DeutscheBank), and members of New York's growing coworking community. The session was an open coworking session with collaborative activities, open discussion and networking. The discussion covered the following issues:

— Current mobile work practices / arrangements in corporate organizations
— The tools, places and organizational policies to support mobile work
— Assessing who are the right workers and what are the right work activities for collaborative work environments
— Lessons corporate organizations learn from co-workers
— Mapping best of / worst of mobile work on colored post it notes

Mobile Office Design
Shake Shack, Madison Square Park, New York City
October 6, 2009
15 attendees took part in a design charrette to envision the mobile office. Participants included architects, interior, furniture and product designers, social scientists and mobile workers/co-workers. The goal of the charrette was to develop designs for collaborative workplaces for mobile workers (indoors or outdoors). By conducting the charrette in a public park, participants faced the challenges and benefits of working in a public environment firsthand. Activities included a discussion of mobile work practices and audience, group brainstorm / design / sketching, and group critique of the proposed designs. The discussion covered issues such as: the audience for mobile work, working in outdoor and semi-public places, and the relationship between a mobile office and existing public space infrastructure. Proposed designs ranged from portable, expandable individual seating, and a wired wall to providing shade and Internet connectivity.

Urban Planning for a Mobile Workforce
Edgar Trinity Plaza, New York City
October 2, 2009
Five attendees took part in this discussion on the role of city government and other public agencies in supporting mobile work. The goal of the collaboration was to generate ideas for public policy or public space planning guidelines to support collaborative, public work environments. Participants included urban planners / designers and workplace strategists. The discussion focused on the following themes:

— How can we make cities more mobile work-friendly?
— Can mobile / flexible work help to activate public places and the local economy?
— What is the value of government support for shared work spaces?
— How can we overcome the assumption that work requires significant physical infrastructure?
— How does Mobile work highlight the importance of location / geography for getting business done?
— How to differentiate the benefits of collaboration from concerns over individual productivity?
— How to make the mental leap from office work to outside work smaller and easier for traditional office workers?

Rethinking How Office Buildings Meet the Street
One of the most significant insights from Breakout! relates to the core assumption of the project, embedded in its name: "escape from the office." We have extensively documented the inspirations for this need to "escape," driven by the shortcomings of modern offices for creative work, new technological freedoms, and the organizational need to connect to outside people and ideas. Yet, as we discovered, while urban public spaces offer many tantalizing potential worksites, the range of challenges to creating conditions suitable for work were much greater than anticipated.

Additionally, the capital investment represented by the existing stock of office buildings is enormous. Thus, while the initial concept of Breakout! sought to challenge and potentially replace the office building as a primary way of structuring work and workspace, one important insight from the project is that public work should inspire us to rethink the relationship between office buildings and public spaces. This rethinking should be focused on developing new techniques and forms to blur the boundary between public workspaces (and public social networks) and private workspaces (and private networks). Here, we outline some of the key issues at hand, and propose a set of guidelines for design that seeks to bridge the divide

Situating work in the park on One Web Day.

An inspirational context for collaborative work.

Urban gardens provide a rewarding place for collaboration, but challenges for using information technology.

between the emerging public knowledge work of the streets and the private workspace of offices by opening up the street level of adjacent office buildings.

The public realm of the city offers many potential workplaces. However, before Breakout!, no clear typology existed for evaluating or comparing public workplaces in the city. In analyzing potential sites for Breakout! events, these sites began to fall into several broad categories that describe how they problematize the urban environment and undermine certain assumptions about the public workplace:

— Places with a favorable context
In Manhattan, events were held at locations such as Washington Square Park and the South Street Seaport. These locations were central to large business districts and a diverse, office-bound workforce with the technological and organizational capacity to work outside the office. Transit vehicles, such as commuter trains and ferries also fall into this category, because they provide opportunities for passersby to serendipitously discover collaborative work sessions and become part of the process.

— Places with an inspirational context
In Manhattan, events were held at locations such as Central Park's Sheep Meadow and Jefferson Market Garden. These sites were intended to provide a mechanism for creating an entirely new context for work, through natural or manmade inspirational settings. These are places that could enhance creative work, but are currently disconnected from our commercial districts or do not provide necessary amenities.

— Re-contextualized places
Looking at Breakout! from an urban design perspective, events were held at the Elevated Acre and Washington Square Village, public spaces dating from the 1950s and 1960s that are largely considered to be under-utilized. The intention was to recontextualize these public venues as workspaces in order to help reactivate them. Ironically, it was their relative isolation from street noise that made them highly attractive places for common knowledge worker tasks like telephone calls and videoconferencing.

Finally, given Manhattan's prominence as a global commercial center, by design many of the Breakout! sites were deliberately selected to be in proximity to a large potential group of participants assigned to traditional office workplaces.

Mobile workers develop many coping mechanisms to create the conditions suitable for their work. Through the Breakout! Mobile Office Kit described in the preceding sections, we hoped to augment the capacity of mobile work teams to make lightweight adaptations and extensions to existing public venues and street furniture. But the interaction between work groups and the different characteristics of the various sites also generated many insights about the obstacles to accommodating mobile work in each.

Three sets of obstacles are most important to address in supporting mobile work.

Activating a poorly used public space through outdoor coworking.

First, the "reptilian" needs of shelter from the elements, connectivity to communications and power grids, and security can be problematic in public spaces. Sunlight in particular makes laptop screens difficult to read, and the power-weight ratio of contemporary batteries means that even a modest team would require a heavy bank of slow-drain boat batteries for a full day's use of phones, laptops, printer and other necessary equipment. While the Breakout! events were able to use smaller batteries to support two–three sessions, longer activities required AC power, which was nearly non-existent in the public spaces we utilized. Wireless connectivity via 3G–Wi-Fi bridge devices was unreliable, and unpredictably slow. Mobile work supported by mobile devices still requires infrastructure, but new kinds of infrastructure that require much less constant tethering, and much more episodic tethering.

Second, environmental conditions in public space can interfere with many of the more common aspects of knowledge work. Laptop computers were largely optimized for indoor "laps," and mobile devices lack the user input and output devices for rich content creation and interaction, although new categories like tablet computers may change this. Ambient noise from traffic, loud conversations, emergency sirens, insects, sudden rain, construction, and a surprising array of civilian and police helicopters were constant obstacles to telephone and videoconferencing. While we understand that Manhattan is an extreme environment, ambient noise was a factor in nearly every event held. The design of public knowledge workplaces in the future will need to inform, leverage, and accommodate a diversity of human-computer interactions, not just personal mobile devices, but also devices designed explicitly for group use.

Third, while Breakout! symbolically drew attention to the potential of public space to serve as an alternative to office space, it didn't explore the connection between the two. Defining how Breakout! sessions and public work were something more than a playful escape from everyday routine was less effective because of the high level of commitment required to participate, as well as the ease of access, and because teams were not gathering on a regular basis. Furthermore, people were drawn from many different organizations, and there is a lack of ongoing collaboration between team members. To an extent, some events just didn't "feel like work". As we discussed earlier, while these events were intended to create intense sites of collaboration, many knowledge workers don't know how to collaborate yet. Places for public knowledge work need to embed this familiarization and education into the very fabric of built space.

For all of these reasons — the need for supportive grid infrastructure, the need for buffers against various environmental factors and the need to make the transition from office to public space more seamless — one of the key insights emerging from Breakout! is to leverage office buildings themselves to provide these functions at their interface to the street and public space. Rather than completely escaping the office building, it is possible to redesign the interface with the street and public space to make it more permeable, and supportive of a diversity of work activities, just as many contemporary designs support a mixture of uses (retail, community, leisure, transit) at ground level. These mobile work-friendly building designs would seek to support a variety of users: tenant workers who want to "break out" of their conventional office space to induce creativity or find opportunities for informal collaboration with other organizations, and creative individuals seeking temporary worksites.

New York's modern commercial architecture is defined by the exclusion of work at the street level, where it is subordinated to the higher rents of retail and dining. Moreover, what sheltered "public" space does exist as lobbies is purposely designed and underfurnished in order to expedite passage and discourage extended dwelling time. While many adjacent plazas and indoor atria are provisioned with seating areas adequate for work (and used in many Breakout! sessions), these are largely intended for very short duration use — for eating, resting or brief conversations, not intense collaborative knowledge work.

In conclusion, as we have discussed, long-term trends in work, connectivity and organization indicates that public spaces are being transformed into knowledge workplaces. This appears to be inevitable, even if the scope of how deeply mobile work will be embedded in public spaces is still an open question, as the challenges and limitations of Breakout! experience suggests. As we move forward in rethinking the interface between office buildings and public space, three principles should be explored:

— Making commercial activity transparent
as organizations evolve and restructure to become more permeable and more transparent, and new highly open forms of organization emerge, these new values should be reflected in the architecture of spaces that support their work. Office buildings solved the 20th century problem of separating, protecting and controlling flows of people, information and knowledge in organizations. Designing an architecture for knowledge work in the 21st century may mean literally blowing open the walls of office buildings, which for high-rises is only really practical at the ground level.

Today, the ground floor of most office buildings is either a blank wall adjacent to a concrete or landscaped foyer, or a place where transparency is reduced to a system of display complementing completely unrelated retailing activities. While we don't expect the New York Stock Exchange to return to the shade of a buttonwood tree, there are numerous opportunities to connect virtual financial networks to public spaces at the foot of office buildings in ways that make those commercial activities more transparent.

— Providing spaces for collaboration
most modern office buildings lack neutral spaces for meetings. Nearly all meeting spaces are provided inside private tenant spaces, or off-site conference facilities. As new models of inter-organizational collaboration and even internal collaboration emerge, there will be an increasing demand for on-site, neutral meeting spaces. As organizations evolve towards more networked structures, the need to engage individuals in many more relationships beyond just employer-employee or contractor-customer ones, will require neutral spaces that are more elaborate and extensive than the local cafe. These spaces will be explicitly engineered to amplify the casual interactions of collaborative organizations.

— Providing landing places for mobile workers
perhaps the most challenging guideline is that office buildings should provide landing places for mobile workers in order to attract valuable people and create the opportunity for fruitful interactions. People who have no official status with building tenants should be allowed to enter the building at these blurred edges. Other kinds of credentials may be imposed, and stacked layers or circles of spaces could be used to filter and accommodate different human needs and resources. The kinds of people these spaces seek to attract can provide a great resource to building tenants by effectively creating an on-demand labor force.

These guidelines are preliminary, but their overall goal is to fold elements of all three types of sites previously discussed into the interface between office buildings and streets. First, to situate knowledge work in a favorable context where many different flows of people converge and many different skills and knowledge sets can be tapped at portals between buildings and the street. Second, to situate knowledge work in an inspirational context, where it can be reconfigured in radically different ways than the existing forms of office, cubicle farm or conference room. Café collaborations demonstrate the power of removing boundaries and schedules from the organization of teams, and park Breakout! sessions put knowledge work in highly stimulating natural settings. Finally, these guidelines seek to encourage recontextualized spaces. The street wall of most office buildings is an underperforming space, but knowledge work can be used to reactivate it while rethinking the standard menu of retail amenities and configurations available today.

These guidelines do not always require massive reconstruction, but can be employed in lightweight ways. Perhaps the most provocative experiment of Breakout! occurred in the street plaza of the Meatpacking District, when a workspace was marked out in an existing public seating and table area. In contrast to many other sessions, passersby clearly understood that the group was working, that they had declared a workplace, and that it was

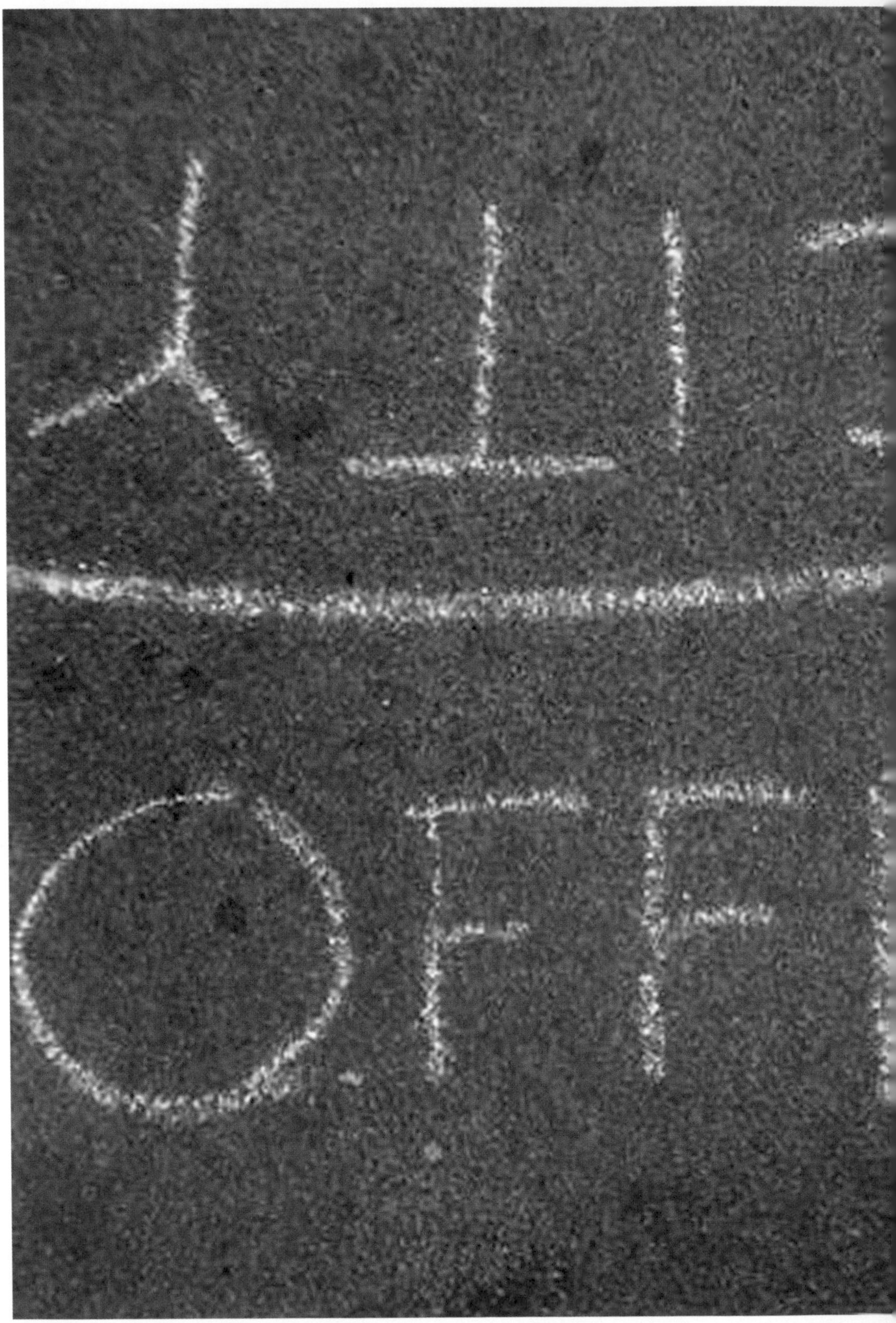

Redefining the relationship between city and workplace can be as simple as drawing a chalk line.

Turning urban plaza seating into an ad hoc workspace.

a well-situated and sensible use of the public space. Going beyond temporary markets and embedding some of these concepts into the interface between buildings and the street is a sensible next step.

QUESTIONS AND METHODS FOR FUTURE EXPLORATIONS OF SITUATED WORK

Researching social, technological and spatial phenomenon such as the nature of situated work in the sentient city is incessantly frustrating. While it is often necessary and productive to hold two opposing viewpoints in mind (Martin, 2007), it is very difficult to maintain focus on three. It is nearly impossible to pull apart the strands of the core research questions for long enough to catch a glimpse of what you might learn if you pursued them. It is fruitful to develop sets of research questions aimed at each aspect of the problem that you are studying while at the same time taking advantage of hybrid concepts that allow for the simultaneous analysis of several factors.

While theoretical constructs such as the social construction of technology (Pinch & Bijker, 1984), actor network theory (Latour, 2005) or the theory of affordances (Norman, 1990) might prove immensely useful, they alone are not sufficient to tackle the complexities of research in this area. Urban informatics, an emergent field of research that is interdisciplinary and includes both conceptual as well as design research, might have some answers, but each passing year contains a great deal more questions due to constant shifts in the socio-technical, economic and political mud in which this topic is embedded.

Methodologically, the practice of designing interventions, along with other forms of inquiry such as statistical analysis, interviews and ethnographic observations, is instructive in learning to think in new ways about the same questions. While this project was conceived more as art than science, one knows not exactly where the line should be drawn and why. Rather, it seems more interesting to push and prod a research question from a wide variety of angles and with an inspired group of people, whether academics or practitioners, until it "cracks," so to speak. Furthermore, research on integrative thinking (VanPatter and Pastor, 2002-2010) that allows for a more holistic understanding of the relationships between people, technologies and places is incredibly useful to furthering theory in this area.

In planning future experiments, interventions, and studies around situated work in the sentient city, several key questions are clear. First, who is responsible for paying and providing the amenities / infrastructure to support work outdoors? Does taking corporate work to public places threaten their "publicness" with privatization? What does the "permeability" of commercial architecture look like and how is it managed?

Designing future public spaces that can accommodate collaborative knowledge work, and contribute to the cultural and commercial well-being of the city will require a broad-based approach involving workplace designers, urban planners, architects and workers themselves. Individuals, corporate organizations and public agencies together must work to answer these questions as the boundaries between office and the city, and between work and life, continue to blur. This project has begun to develop a vocabulary for new ways of working and living that span and dissolve the dichotomy between private commercial space and public civic space. These tools and materials are available to the public under a Creative Commons license at breakoutestival.org.

ADDITIONAL PROJECT CONTRIBUTORS

Dana Spiegel, Amanda Kross (DEGW), Jung Hoon Kim (DEGW), Georgia Borden (DEGW), Sean Savage, Ramon Sanguesa (Citilab Cornellà)

REFERENCES

— Benkler, Y. (2006). The Wealth of Networks: How Social Production Transforms Markets and Freedom: Yale University Press New Haven, CT, USA.
— Castells, M. (1996). The Rise of the Network Society. Malden, MA: Blackwell Publishers.
— Duffy, F. (2008). Work and the City. London: Black Dog Publishing.
— Forlano, L. (2008). When Code Meets Place: Collaboration and Innovation at WiFi Hotspots. Columbia University, New York.
— Forlano, L. (2009a). WiFi Geographies: When Code Meets Place. The Information Society, 25, 1-9.
— Forlano, L. (2009b). Work and the Open Source City. Urban Omnibus, 2009
— Girard, M., & Stark, D. (2002). Distributing Intelligence and Organizing Diversity in New Media Projects. Environment and Planning A, 34(11), 1927-1949.
— Gottmann, J. (1983). The Coming of the Transactional City. College Park, MD: University of Maryland.
— Hennion, A. (2004). Pragmatics of Taste. In M. Jacobs & N. Hanrahan (Eds.), The Blackwell Companion to the Sociology of Culture. Malden, MA: Blackwell.
— Humphreys. (2008). Mobile Social Networks and Social Practice: A Case Study of Dodgeball. Journal of Computer-Mediated Communication, 13(1), 341-360.
— Latour, B. (2005). Reassembling the Social: an introduction to actor-network-theory. Oxford: Oxford University Press.
— Martin, R. L. (2007). The Opposable Mind. Boston, MA: Harvard Business School Press.
— Norman, D. A. (1990). The Design of Everyday Things. New York: Doubleday.
— Oldenburg, E. (1999). The Great Good Place: Cafes, Coffee Shops, Bookstores, Bars, Hair Salons, and Other Hangouts at the Heart of a Community. Cambridge, MA: Da Capo Press.
— Pinch, T. J., & Bijker, W. E. (1984). The Social Construction of Facts and Artefacts: Or How the Sociology of Science and the Sociology of Technology Might Benefit Each Other. Social Studies of Science, 14(3), 399-441.
— Powell, A. (2009). WiFi Publics: Producing Community and Technology. Information, Communication & Society, Vol. 11, No. 8, pp. 1068-1088, December 2008.
— Rheingold, H. (2003). Smart Mobs. Cambridge, MA: Perseus Books Group.
— Sarbaugh-Thompson, M., & Feldman, M. S. (1998). Electronic Mail and Organizational Communication: Does Saying" Hi" Really Matter? Organization Science, 9(6), 685-698.
— Sobel, R. (2000) Curbstone Brokers: The Origins of the American Stock Exchange. Beard Books.
— Tripsas, M., & Shah, S. K. (2007). The Accidental Entrepreneur: The Emergent and Collective Process of User Entrepreneurship. Harvard Business School Entrepreneurial Management Working Paper No. 04-054.
— Uzzi, B., & Dunlap, S. (2005). How to Build Your Network. Harvard Business Review, 83(12), 53-60.
— USGBC. (2010). Green Building Facts. Washington, DC: U.S. Green Building Council.
— VanPatter, G., & Pastor, E. (2002-2010). Design Thinking Made Visible Project, Integrative Thinking Research Initiative: Humantific Inc.
— Von Hippel, E. (1978). Users as Innovators. Technology Review, 80(3), 31-39.
— Von Hippel, E. (2005). Democratizing Innovation. Cambridge: MIT Press.

ESSAYS

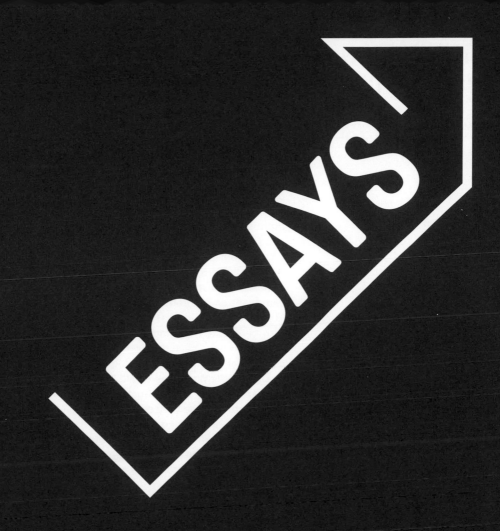

THE ACTION IS THE FORM

KELLER EASTERLING

"...tools only exist in relation to the intermingling they make possible or that make them possible." [1]

Digital infrastructure is just one of the things that, in its ubiquity, often becomes more obscure. Infrastructural space is, as the word suggests, customarily regarded as a hidden substrate — the binding medium or current between objects of positive consequence, shape and law, yet it is also the point of contact and access, the spatial outcropping of underlying laws and logics. The pools of microwaves that bounce from satellites or the thickening tangles of fiberoptic submarine cable that lie on the bottom of the ocean, however invisible, nevertheless materialize in everything from atomized swarms of electronic devices to building materials and fixtures of urban public space. Moreover, some infrastructural formulations seem to make manifest and press into view a hyperbolic cartoon of their abstract technical and economic logics. Repeatable formulas for spatial products like resorts, malls, IT campuses or free zones manifest in gigantic world city formations. The building enclosures typically considered to be geometrical, formal objects receiving transportation, communication and utility networks have themselves become infrastructural—physical, spatial media and technologies moving around the world as repeatable phenomena. No longer simply what is hidden or beneath another urban structure, many infrastructures are the urban formula, the very parameters of global urbanism.

With minds trained to name and declare, we parse the world with a nominative habit of mind in which nouns are things that can be known and verbs are things that move. Digital architecture, a function of processes, protocols and topologies, is just the most recent mode of exchange to rehearse a currency in shared processes. Yet, while these tune the imagination to the infinitive rather than the nominative, the discipline of architecture still maintains its primary currency in singular durable objects that can be framed and represented. Even in a broader culture, it is likely that most do not look at a concrete highway system and perceive agency. Agency in networks might only be assigned to the moving cars on the road, the electrical impulses in the fiber or the swooshing sound of the sent email. Things that are not moving or expressing their dynamism in some way are not active. They are not *doing anything*.

Borrowing from Marshall McLuhan, the nominative and kinetic both act as *"the juicy piece of meat carried by the burglar to distract the watchdog of the mind."*[2] Borrowing from philosopher Gilbert Ryle, the possibility that inert non-human objects have agency becomes a "ghost in the machine," or borrowing from Jacques Rancière, such a conception becomes "inadmissible" evidence in dominant cultural epistemes.[3]

For architecture and urbanism, as for many schools of thought, the distinction between understanding form as object and form as action is something like Ryle's distinction between "knowing that" and "knowing how." He provides a clown's performance as an example. "Knowing how," like knowing how to be funny, is not something that can be declared or named or reified as an object or event. It is for Ryle, "dispositional."[4] Ryle enjoys the ways in which dispositional expressions thrive in common parlance

1. Gilles Deleuze and Félix Guattari, A Thousand Plateaus : Capitalism and Schizophrenia (London: Athlone Press, 1988). Quoted in Nigel Thrift, Spatial Formations (London: Sage Publications, 1996), 264.
2. Marshall McLuhan: Understanding Media: The Extensions of Man (New York: McGraw-Hill; London: Routledge & Kegan Paul, 1964., 2001), 19. The quotation: "For the "content "of the medium is like the juicy piece of meat carried by the burglar to distract the watchdog of the mind."
3. Jacques Rancière, The Politics of Aesthetics : The Distribution of the Sensible (London ; New York: Continuum, 2004), 85.
4. Gilbert Ryle, The Concept of Mind (Chicago: University of Chicago Press, 1949), 27-32, 43, 89, 116, 119-120.

and are used as a way of describing an unfolding relationship of potential, relative position, tendency, temperament or property in either beings or objects. François Jullien has given the example of a round ball and an inclined plane as a situation possessing disposition—the potentials of a situation as they are associated with factors including geometry and position among many other things. [5] But the ball does not have to move or roll down the hill to possess this disposition. Disposition is composed of sequential action. Ryle emphasizes the latency and indeterminacy of this dispositional action in both human and non-human subjects. A person has the capacity or tendency to sing or smoke. A dog can swim. Rubber loses its elasticity. Glass is brittle. A clown is funny. In this way, Ryle demonstrates that seemingly inert objects are actors possessing agency. They are doing something. Ryle finds great sport in noting that while we work with dispositional expressions in everyday speech, in some logical systems this latent activity is treated as a fuzzy imponderable or an occult agency in "a sort of limbo world."[6]

Infrastructure, whether composed of digital, building or urban components is dispositional. It is made of action just as much as it is made of concrete, bits, cables or CPUs. It does not constitute an event, but must rather be observed over time as a potentiality, capacity, ability, or tendency. Its activity is not reliant on movement but rather on unfolding relationships inherent in its arrangement. Designing infrastructure is designing action. The contemplation of disposition tutors these artistic faculties, unused in some disciplines and a staple in others. With highly developed discourses to treat object, content, outline and nominative, architecture and urbanism remains under-rehearsed in making action, medium, relation, or infinitive and may even regard the possibility of active form as oxymoronic.

If making action is not a recognized artistic faculty, one would need to inform the clown in Ryle's example. Indeed, the transposition from the nominative to the active that requires so much ideation and analysis in some schools of thought, like design, is a completely ordinary or practical matter in some other disciplines like theater. Working up to their elbows in the construction of dispositional action, those in the theater come very close to handling action as an essential raw material. An actor adheres to an explicit script, but the scripted words are only considered to be traces or artifacts that provide hints of an underlying action. An actor constructs a scene as a string of sequenced actions. Often it is that action that is the meaning or information conveyed. Actors rarely deal with nominative or descriptive expressions—states of being or mood. One cannot play "being a mother," for instance. Because it is fixed and nominative, this is usually a bad performance that lessens the possibility of listening to and interacting with other performers — a form of over articulation known as "indicating" because its self-reflexivity lessens the possibility of listening to and interacting with other performers. Theatrical techniques often privilege infinitive active expressions. The director asks the actor, "what are you doing?" Letting a vivid action carry the words rather than the other way around is a relatively durable technique. It is the action or driving intent, referenced as an infinitive expression, that is leading the performance—not

5. Francois Jullien, The Propensity of Things: Toward a History of Efficacy in China (New York: Zone Books, 1995), 29.
6. Ibid.,Ryle, 119-120.

movement, gestures, blocking or choreography. [7] Since motherhood is an abstraction, an actor would play not "being a mother" but rather "smothering a child." Not the text, but the action, is the real carrier of information.

The notion that social and technical or socio-technical networks like infrastructure are *performing* is one that Bruno Latour has long posited in his renovation of social "science."[8] Both Ryle and Latour enjoy holding up the artifacts that do not fit into the box or the butterflies not pinned to the board. While many of those studying socio-technical networks were focused on the way social constructs shaped technology, Latour's more radical inquiry considers not only humans, but also technologies as actors. For instance, highways, the electrical grid or a computer are active non-human agents influencing the desires of social networks that reciprocally shape them. Rather than "placeholders" that reinforce existing assumptions, things, whether they are human or non-human, have agency; they are actively "doing something." [9] Latour calls attention to an unfolding trajectory of activities between humans and nonhumans that is harder to fix. Action, he writes is "dislocated" or indeterminate. It is "borrowed, distributed, suggested, influence dominated, betrayed, translated." To study social networks is to continually "follow the actors." [10] Latour writes:

> *It is not by accident that this expression, like that of 'person', comes from the stage.... To use the word 'actor' means that it's never clear who and what is acting when we act since an actor on stage is never alone in acting. Play-acting puts us immediately into a thick imbroglio where the question of who is carrying out the action has become unfathomable.* [11]

If infrastructural organizations are performing, what are they doing? If their performance is indeterminate, how are they designed? As impossible as these concepts may seem within some disciplinary logics, contemplating dispositional activity opens onto a fresh field of endeavor. For designers and urbanists, such a contemplation redoubles our form-making capacities to include active forms — spatial agents or actors that shape not only the objects, but the way the object plays — what the object is doing. They condition material and immaterial parameters, aesthetic practices and political trajectories. For instance, active forms may describe the way that some alteration performs within a group, multiplies across a field, reconditions a population or generates a network. They may be not only physical objects or contagions, but also topologies or organizational properties within a spatial field. The designer of active forms is designing the delta or the means by which the organization changes — not the field in its entirety, but the way it is inflected, the dispositions immanent within its organization. So, while perhaps intensely involved with material and geometry, active forms are inclusive of, but not limited to enclosure and may move beyond the conventional architectural site. Active forms are not at odds with, but rather propel, expand, (even rescue) form as object. As they may ride larger organizations, they offer additional modes of authorship with time-released powers and cascading effects.

7. Sharing a sensibility with theater, Ryle, for instance, makes as distinction between active verbs or "performance verbs" and verbs like "'know,' 'possess' and 'aspire.'" One would not say, for instance "'he is now engaged in possessing a bicycle.'" Gilbert Ryle, The Concept of Mind (Chicago: University of Chicago Press, 1949), 130, 116. 8. Bruno Latour, Reassembling the Social: An Introduction to Actor-Network Theory (Oxford: Oxford University Press 2005), 5, 8-9, 10-11.
9. Ibid., 46.
10. Ibid., 46, 39.
11. Ibid., 46.

While perhaps initially obscure, the idea that static objects and organizations have agency is only a discovery of something we knew all along, just like we know there is no way to answer the question, "What is funny?" The urban environment grows or changes because of active forms within it, whether they be contagions or topologies. For instance, an elevator, spatial product, law, real estate wrinkle, financial formula, network topology, material imperative, or persuasion may be an active form within the city. They may be designed with immaterial parameters that may only have eventual material consequence. It is of little consequence to alter one house in a suburban field, but it is very effective to design a real estate protocol that is contagious within it. It is relatively meaningless to attempt to represent a process like the Internet, but very meaningful to author active forms that ride that network. Similarly, one cannot design diversity in a city by crafting variability in its individual components, but one can design an urban infrastructure from both geometry and relationship that continues to generate diversity, and is reliant on both the shape of physical form and the scripts that govern their use and growth. These combinations of form and protocol contribute deliberate tools for adjusting organizational constitution, and they are capable of rendering mixtures that are, for instance, homogeneous, heterogeneous, monopolistic, oligarchic, open, resilient or recursive.

Gregory Bateson's adventurous thinking, taken together with that of Ryle and Latour, further delineates how one might begin to inflect political disposition and even temperament immanent in organization. Just as Ryle and Latour see no separation between human and non-human actors, Bateson addresses a world made of everything, not a world subdivided into the subjects of different sciences. He speculates about activity embedded in organizations made of individuals or sounds or circuits or neurons. As a cybernetician, Bateson characterizes information as a universal unit or elementary particle. "Information is a difference that makes a difference," he famously wrote. [12] Objects as well as actions are not anthropomorphized as little selves that possess mood and intentionality, but the degree to which they "make a difference" in the world constitutes influence, intention or *information*.

Information shapes morphology and organization in biological or machinic, human or non-human, systems. Assessing any group — electronic circuits, nations, tribes from New Guinea or Alcoholics Anonymous meetings — with this cybernetic epistemology, Bateson could also transpose sociological assessments of tension and violence to organizations of inanimate objects. Where Ryle describes disposition as inherent properties (e.g. glass that is brittle), Batson can naturally extend an understanding of disposition to include behaviors inherent in groups. He speculated about the violence inherent in binaries, the way in which that violence might escalate as the binaries become more symmetrical, and the way in which it might be relieved by reciprocal or cooperative activity among multiple power centers. Bateson linked information flow in organizations to dispositions of productivity, stability, violence and collapse. In the competitive or destructive states, the flow of information collapses, whereas in more balanced arrangements information is more easily exchanged. Setting aside some holistic conclusions and

12. Gregory Bateson, Steps to an Ecology of Mind (Chicago: University of Chicago Press, 2000), 381, 462, 315, 272, 21.

codifications of cybernetics, Bateson's simple speculations foster an understanding of stability, tension, violence, aggression, interdependence or competition that are literally immanent in urban organizations.

Again, while it might seem odd to speculate that non-human organizations have temperament over and above the human agency within them, the idea finally only returns to something we already know. A term like disposition perhaps only brings the familiar into focus. For instance, two warring factions marching towards each other are symmetrically arranged in a way that fuels violence. An underground mafia organized as a hub and spoke organization fosters secrecy because of limited contact to administrative decisions. A television or radio organization of mass media similarly has a hub and spoke organization very different from contemporary networks of computation. A skyscraper organizes sequential movement not unlike serial computing. A mat building with multiple points of entry is something like its parallel computing counterpart. A telecom locates its underground fiberoptic cable in relation to only one segment of the population, and it operates as a monopoly. In each of these examples the active forms or directions for activity have a substrate of geometry or arrangement that shapes the disposition of the organization. Each of these topologies or relative power positions possesses a quotient of, for instance, violence, resilience, competition, patency or closure.

The contemplation of disposition also tutors political faculties. The most powerful players have the capacity to make infrastructure, but equally important, infrastructure can escape nominative designations or documented events. As action, it can remain undeclared and discrepant, and, as medium, it can determine what survives. Different from the politics that names and squares off against every opponent or tries to kill every weed in the field, the indeterminate dispositional space of infrastructure may neutralize or adjust by changing the chemistry of the soil. The broad foundational transformations of infrastructure change, like sea changes or changes to an operating system, offer a special political instrumentality that may preclude the fight. While those political traditions that call for inversions and revolutions often call for the absolute annihilation of the preceding system, lateral techniques of dissensus work on the *ongoing* reconditioning of a spatio-political climate.

The projects commissioned for the Sentient City exhibition embed digital technologies that we recognize as active into urban surfaces that we customarily consider to be inert. Unlike those digital installations that signal technological anthropomorphism or dynamism with an animation of blinks and beeps, the installations are often heightening an awareness of either a lurking digital technology or a relational agency existing in the urban environment. At their best, while they maintain independence as techno-artistic urban performances, they highlight some of the same territory that Ryle, Latour and Bateson find on the other side of an altered habit of mind. It is territory where the action is the form.

INTERACTION ANXIETIES

OMAR KHAN

In a recent article[1], interaction design consultant Donald Norman sounded some warnings about the new trend in interaction design — the *natural user interface* (NUI). NUIs look to replace the graphical user interface (GUI) with more "natural" interactions including speech, touch and gestures. Steve Ballmer, CEO of Microsoft, is quoted in the article as saying that 2010 will be remembered as the year when the shift to NUIs took place. Norman is not convinced. He takes exception to the "natural" designation of NUIs and warns against the limits of gestures for interaction design. His reservations include that gestures are not natural, but like graphical interfaces have to be learned. They are ephemeral and don't leave a trace of their path, thus providing little feedback to users, and that they can easily be misinterpreted by people and more importantly computers. Norman's critique however, is tempered by his recognition that NUIs "will enhance our control, our feeling of control and empowerment, our convenience and even our delight."[2] But that will only come once NUI's develop "well-defined modes of expression, a clear conceptual model of the way they interact with the system, their consequences, and means of navigating unintended consequences."[3]

Norman's objections must be placed in the context of interaction design's historical focus on the workplace machine.[4] Here research has pursued the design of effective interfaces, hardware and software through which information in a computer's memory can be easily accessed and manipulated. Its products — the mouse, keyboard, stylus and GUI — have transformed computers from specialized machines to universal work appliances. Norman's skepticism reflects the limits of NUIs for the types of interactions that we have become accustomed to with our GUIs. It also reflects a deeper anxiety with the changing nature of computing that is increasingly mobile, materially embedded and pervasive. Perhaps the workplace machine is not a suitable model for this type of interaction? Or that effective interfacing is the measure of effective or affective interaction in the age of pervasive computing? Could it be that interactions will not only be for information exchange but designing, provoking and situating a variety of social and cultural practices? With buildings, clothes, objects and places becoming computationally augmented we need to take a more holistic view of interactivity and explore how it can assist in constructing productive and provocative relations between people, places and computing instruments. What role does space, mobility and embodiment play in such constructions? How will interaction affect our understanding of our own agency in perceiving and acting in space? And what of the agency of sentient systems through and with which we will interact?

The expansion of our understanding of what interactivity could be as computation becomes pervasive requires a shift away from the instruments of interactions — screens, mice, speech, gestures, tangible interfaces — and towards the relations we expect to achieve from them. These include the ways in which we communicate and socialize with one another and inhabit our cities and world. We need to speculate on the

1. Donald Norman, "Natural User Interfaces are not natural," Interactions, v.XVII.3 (2010): 6-10.
2. Ibid., 6.
3. Ibid., 9-10.
4. See Bill Moggeridge, Designing Interactions, (Cambridge, MA: MIT Press, 2007) which is a personal chronicling the development of computer interfaces.

cultural and aesthetic worth of interactivity in order to accommodate it more properly in our lifestyles. At the same time, we also need to recognize the opportunities that computing in its different forms — mobile, embedded, and pervasive — offers for changing our expectations and usage of space, architecture and urbanism. Interactivity's unique aesthetic potential for our media, architecture and cities requires the participation of designers, artists, architects and urbanists to help situate these technologies. With pervasive computing's technological inevitability it is imperative that designers, architects and artists contribute to the imaginary of these sentient systems.

Interactivity's Destabilizing Aesthetics

In the early theorizing of interactive art, the integral role of the viewer as an active participant in the construction of the aesthetic experience was noted. [5] Burnham observed that interactivity's two-way communication between observer and artwork resulted in a "figure-ground reversal in human perception of the environment."[6] This resulted in an aesthetic shift from a fixed viewer-object relationship to one in which the observer was understood as an integral part of his or her environment. Further, through interaction the separation of the viewer and the work of art was negated, "fusing both observer and observed, 'inside' and 'outside'."[7] Hence, as interactivity empowers the observer to engage and influence the work of art, it destabilizes this control by allowing her to lose herself through the work. In other words, interaction puts the observer at risk, such that her participation can result in desired outcomes or unpredictable surprises or even utter failures. This is an important aspect of interactivity's aesthetic effect.

This is different from Umberto Eco's observation of the aesthetics of the "open work."[8] Eco's polemical study of modernist works in which the performer or reader is tasked with "finishing" the work through his/her "reading" suggests a similar aesthetic engagement. However, the open work engages the reader in a more structured way. Eco explains, "The author offers the interpreter, the performer, the addressee, a work to be completed. He does not know the exact fashion in which his work will be concluded, but he is aware that once completed the work in question will be his own. It will not be a different work, and, at the end of the interpretive dialogue, a form which is his form will have been organized, even though it may have been assembled by an outside party in a particular way that he could not have foreseen."[9] Hence the reader remains "outside" the artwork. Where the open work requires interpretation from the observer, interactivity requires intervention.[10]

Espen Aarseth in his study of cybertexts calls this a "cyborg aesthetic."[11] Taking his cue from Donna Harroway's "A Cyborg Manifesto" (1991) who used the cyborg, "a hybrid of machine and organism," as a concept to challenge fixed categories of gender, nature, race and identity, Aarseth speculates on the cyborg as a means to problematize power and control structures. He writes, "Any cyborg field, as any communicative field, is dominated by the issue of domination and control. The key question in cyborg aesthetics is therefore, who or what controls

5. Jack W. Burnham, "The Aesthetics of Intelligent Systems," in On the Future of Art, ed. Edward F. Fry (New York: The Viking Press, 1970), 95-122.
6. Ibid., 100.
7. Ibid., 103.
8. Umberto Eco, The Open Work, trans. Anna Cancogni (Cambridge, MA: Harvard University Press, 1989).
9. Ibid., 19.
10. Espen J. Aarseth, Cybertext-Perspectives on Ergodic Literature (Baltimore: Johns Hopkins University Press, 1997).
11. Ibid., 54-56.

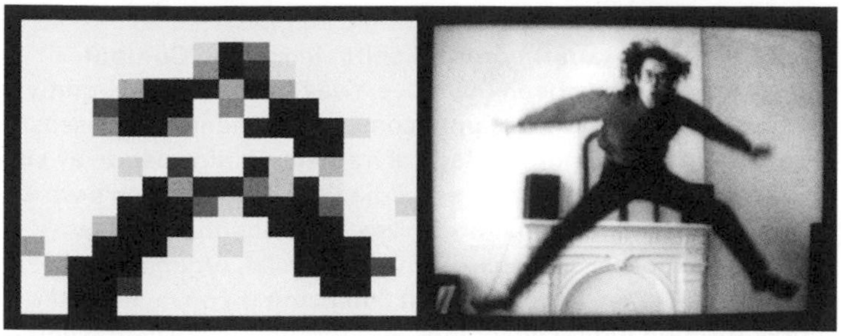

David Rokeby's "Very Nervous System" (1986-1990), demonstrates the aesthetics of losing oneself to the work, a key characteristic of interactivity. Image courtesy David Rokeby.

the text? Ideologically there are three positions in this struggle: author control, text control and reader control."[12] And then following a discussion of John Cayley's computer program, Book Unbound (1995) that algorithmically produces text through user interaction:

the text is an impurity, a site of struggle between medium, sign and operator. The fragments produced are clearly not authored by anyone. They are pulverized and reconnected echoes of meaning, and the meaning that can be made from them is not the meaning that once existed... The pleasure of this text is far from accidental; it belongs not to the illusion of control but to the suggestive reality of unique and unrepeatable signification.[13]

To further explicate this, David Rokeby's Very Nervous System (1986-1990) is a powerful demonstration of a cyborg aesthetic. Rokeby designed a machine vision system that could interpret physical gestures and translate them into sounds. Moving your arms and legs at different speeds and heights produced commensurate sounds that you could come to control or give yourself up to. As Rokeby explains, "The feedback is not simply 'negative' or 'positive', inhibitory or reinforcing; the loop is subject to constant transformation as the elements, human and computer, change in response to each other. The two interpenetrate, until the notion of control is lost and the relationship becomes encounter and involvement. The diffuse, parallel nature of the interaction and the intensity of the interactive feedback loop can produce a state that is almost shamanistic. The self expands (and loses itself) to fill the installation environment, and by implication the world."[14]

The simultaneously enabling, yet destabilizing, nature of interactivity undermines who or what is in control. Instead, interactivity puts control into play, something to be negotiated in the performance of the act. A genuine dialogue or conversation with the computer remains the ideal [15], but in lieu of it we are witnessing that even lopsided two way communications, like those we have with our pets, yield provocative, but extremely fulfilling exchanges. While there is a palpable anxiety that results from the unanticipated barking of such exchanges, we can look forward to more intimate and unpredictable relationships with our architecture and places of habitation.

12. Ibid., 55.
13. Ibid. 56.
14. See David Rokeby's website for Very Nervous System: David Rokeby, "Very Nervous System," http://homepage.mac.com/davidrokeby/vns.html.
15. A computer's ability to carry on a dialogue with its user remains the measure of artificial intelligence. Alan Turing's famous Turing test, as proposed in his paper "Computing Machinery and Intelligence", Mind LIX 236 (1950): 433–460, made the conversational exchange an indicator of intelligence. John Searle's "Minds, Brains and Programs", Behavioral and Brain Sciences 3 (3) (1980): 417–457, counters that an intelligent conversation could be simulated by a machine indicating neither understanding nor a consciousness. See also cybernetican Gordon Pask's conversation theory that lays out axioms for conversational interactivity with machines: Gordon Pask, Conversation Cognition and Learning (Amsterdam: Elsevier, 1975).

Cybernetic Organizations

One of the important theoretical shifts in Human Computer Interaction (HCI) has been the move away from a naive cognitivist view of information as a pure construct of mental processes towards a more phenomenological framing of information as situated [16], contextual [17] and embodied [18]. Paul Dourish's work on embodied interactions recognizes "that action and meaning arise in specific settings — physical, social, organizational, cultural, and so forth,"[19] and that "meaning is conveyed not simply through digital encodings, but through the way that computation enlivens those encodings with semantic and effective power." [20] For interactivity this suggests that our lived context is also in play in human-computer communications. This should be encouraging for architects, designers and artists, for whom the "context" of everyday life is a familiar part of their work.

One caution that I have with this approach is that it runs the risks of reifying the everyday. With its focus on developing interactive technologies that support situated actions, it inadvertently fixes located practices and undermines any technology that might accidentally or intentionally disrupt them. As Malcolm McCullough explains, "contexts remind people and their devices how to behave. That framing has often been done best and understood most easily as architecture. Something about the habitual nature of an environmental usage gives it life. Like device protocols and personal conduct, architecture has the form of etiquette. Like most etiquette, architecture exists not out of pompousness, but because it lets life proceed more easily. Situated computing extends this age-old preference, where as anytime-anyplace computing does not."[21] It is unclear to me whether truly interactive systems can subscribe to any preordained etiquettes. A cyborg aesthetic precludes such notions of "good" and "bad" behavior since meanings emerge out of the interaction. As such, it puts us in a participatory relationship with our environment, fortifying our agency while not necessarily reciprocating it with control.

Perhaps another way to address the issue of "context" is rather than fixing it to also put it in play with the interacting subject and computing technologies. This would suggest a more complex ensemble of interacting parts that includes people, places and things in a feedback loop with one another. It would also require that we take the cybernetic aspect of the cyborg aesthetic more seriously. The question to ask is, what kinds of relationships would emerge out of such feedback systems? And how does context also become a product of the interaction? In "What is Interaction? Are there different types?"[22] Dubberly, Pangaro and Haque take a pass at this by developing an expanded taxonomy of interaction. They describe interactive systems as a set of relations — linear, self-regulating and learning — that can be combined with one another to create more complex interactions. Theirs is a systems design approach, where the behaviors, rather than the particular mechanism that produce them, are explored.

Linear systems are those that take inputs and produce some predictable output. They demonstrate a cause and effect relationship which makes them more reactive than truly interactive.

16. Lucy Suchman, Plans and Situated Actions: The Problem of Human-Machine Communication (Cambridge, UK: Cambridge University Press, 1987).
17. Malcolm McCullough, Digital Ground–Architecture, Pervasive Computing and Environmental Knowing (Cambridge: MIT Press, 2004).
18. Paul Dourish, Where the Action Is- The Foundations of Embodied Interaction (Cambridge, MA: MIT Press, 2004).
19. Ibid., 161.
20. Ibid., 163.
21. Malcolm McCullough, Digital Ground- Architecture, Pervasive Computing and Environmental Knowing (Cambridge: MIT Press, 2004), 118.
22. Hugh Dubberly and Paul Pangaro and Usman Haque, "What is Interaction? Are there different types?" Interactions v.XVI.1 (2009): 69-7.

We witness this with our computers when we query them to perform some information processing or when an automatic door opens in response to our proximity. Self-regulating systems are systems with goals that establish cyclic or feedback relationships with different constituents. For example, a thermostat in a room set to 70 degrees performs the function of turning a boiler on or off depending on the room's temperature, which in turn activates the thermostat. The function and response of each component — thermostat, boiler, and room temperature — mutually regulate the comfort of the room. In cybernetic terms, control is distributed since all the parts affect one another. Finally, there are learning systems that are able to evaluate their goals and change them in order to follow new ones. This is the realm of sentient systems that can learn from their context and change their actions accordingly. Humans are of course extremely complex learning systems capable of adapting their intentions at will. But this is in response to specific contexts, situations and environments. Hence, as a learning system they are not autonomous, but part of an ensemble of interacting parts that include other people, technologies and environments.

Let me explain this through an example. During the hot summers of New York City it is common to see an open fire hydrant used by local residents to cool themselves and the street. This is illegal, as it compromises the hydrant's function for firefighting. To address this, the city has put water saving spray caps on the hydrants that allow them be used for recreational purposes, but with limits. Supervision of the technology remains in the hands of the fire department which more often than not results in residents clandestinely removing the caps. The problem of the hydrant is that it is tasked with two very different programs: one, fire safety, and the other, recreation (although public health could also be included in this). The cap solution is purely technological and hence linear. The fire department is responsible for legislating both safety and recreation. If one was to frame the solution as self-regulating, then the users of the technology would have to be included in regulating and maintaining the technology. This would require a different governance structure involving block clubs, resident education and a more user friendly cap. Finally, if we were to imagine it as a learning system, it would include an information layer that could anticipate changing needs based on weather, water pressure, local fires, past use, et cetera, and condition future use of the technology accordingly. More significantly, such a development would make residents, the fire department and other governing bodies more aware of one another and their role in maintaining the health of their environment.

What is helpful about a systems approach to interactions is that we are in the realm of modeling — where we can speculate on the behaviors of interactions without getting bogged down by their specific mechanisms. This is a fundamental part of designing, where options can be imagined and studied before they are subjected to the necessary rigors of problem solving for a specific context. Modeling should not blind us to situational realities but likewise shouldn't be censored by them. Also, such a structure maintains the indeterminacy of interactivity, with the different actors in play exercising different controls on one another. The formulation of these three particular types — linear, self-regulating and learning is not incidental, but comes out of the history of cybernetics [23] where inquiry into "control and communication in the animal and machine," [24] recognized that particular behaviors like self-regulation and learning could emerge out of the cyclic communication witnessed in interactivity. As such, they reflect a historical effort in coming to terms with responsive behavior across biological, social and technological systems. [25]

Shifting Agencies

As our devices, buildings and cities become interactive, we will need to address the possibility that they may require little or no involvement from us to carry out their functions. The evolving Internet of Things envisions connecting a host of heterogeneous digital devices through Web 2.0 protocols to create self-directed communications between such objects. "From anytime, anyplace connectivity for anyone, we will now have connectivity for anything."[26] Digitally enabled things will autonomously produce (sense and process) information locally and share it globally with other devices. One can imagine a self-regulating city where buildings monitor their own energy resources, negotiate their needs with a smart energy grid, and communicate with other buildings to better collectively manage their shared resources. Human participation in the exchange would be minimal since the interacting systems would be well programmed in bartering with one another. What does this say about interactivity? For one thing, it would suggest that human participation need not be central to it. Interacting buildings at a minimum could qualify as self-regulating and at their most ambitious capable of learning. But can interaction take place without human involvement or, at the least, is human observation a necessary part of it? How are we to understand our own agency in this coming Internet of Things and the Sentient City?

One way to understand our relationship with sentient things is through *automation*, where machines are tasked with performing work in our place. The imaginary of the robot is perhaps the most appropriate example of this, which as a human proxy performs the rote tasks that we find too tedious or complicated to do ourselves. But does automation preclude a role for humans in the process? Lev Manovich argues that in automation human involvement moves from active to passive. He writes, "It is important to note that automation does not lead to the replacement of human by machine. Rather, the worker's role becomes one of monitoring and regulation: watching displays, analyzing incoming information, making decisions and operating controls."[27] In other words, interactivity is qualified with a requirement to wait for something to happen before there is a need to act. As such delegating one's active agency to perform some task to a sentient machine does not preclude human participation, even if it is only observation; it only defers it.

Another way to think about our shifting agency with regard to the Internet of Things is through *interpassivity*, which is the "uncanny double" of interactivity. As Slavoj Zizek explains, "The

23. For a succinct general review of the history of cybernetics (first and second order) see Bernard Scott, "Second-order cybernetics: an historical introduction," Kybernetes, Vol. 33 No. 9/10 (2004): 1365-1378. For a more in-depth reading of the history of cybernetics see Steve J. Heims, The Cybernetics Group (Cambridge MA: MIT Press, 1991) and the transactions of the Josiah Macy Conferences on Cybernetics 1950-54: Heinz von Foerster, ed., Cybernetics, circular causal and feedback mechanisms in biological and social systems: transactions (New York: Josiah Macy, Jr. Foundation, 1952) .
24. Norbert Weiner's subtitle to his polemical book on cybernetics: Norbert Weiner, Cybernetics: Or Control and Communication in the Animal and the Machine (Cambridge, MA: MIT Press, 1948).
25. See the transactions of the Josiah Macy Conferences on Cybernetics 1950-54: Heinz von Foerster, ed., Cybernetics, circular causal and feedback mechanisms in biological and social systems: transactions (New York: Josiah Macy, Jr. Foundation, 1952).
26. International Telecommuncations Union, The Internet of Things-Executive Summary, (Geneva: International Telecommunication Union (ITU), 2005): 2.
27. Quoted in Erkki Huhtamo, "From Cybernation to Interaction: A Contribution to the Archaeology of Interactivity," in The Digital Dialectic, ed. Peter Lunenfeld (Cambridge, MA: MIT Press, 2000), 96-110.

A wind activated prayer wheel that prays for you. Photo Thomas Olsson.

obverse of interacting with the object (instead of just passively following the show) is the situation in which the object itself takes from me, deprives me of, my own passivity, so that it is the object itself which enjoys the show instead of me, relieving me of the duty to enjoy myself."[28] Zizek gives the example of the Tibetan prayer wheel that mechanically turns to perform prayers. The worshipper can activate it or more practically let the wind turn it to do the praying for him. Other examples include the chorus in a Greek tragedy that feels for you, the canned laughter on a TV-comedy sitcom that laughs for you, and the movie recording VCR that watches the movie for you. Unlike the interactive or the automated where the human subject is active (in different degrees) through the other objects, in the interpassive the subject forgoes participation, even the passivity of observing, and draws pleasure from delegating that passivity to the objects.

This is the guilty pleasure that we hope for from "smart grids", "self-regulating buildings" and "smart materials." They forecast solutions for the pressing problems of climate change and sustainable energy without requiring any substantial activity from us. We won't need to curb our own consumption nor examine the way we do things, since our sentient buildings will manage the problem. More importantly, they will handle our guilt by hiding our responsibility from us. As one of the provocative images of ubiquitous computing suggests, "they weave themselves into the fabric of everyday life until they are indistinguishable from it."[29] But, at the least, the pleasure will still be ours.

Conclusion

If gestures and speech are the next big thing in the silicon alleys and valleys of the world then that is a good development. At least we will be using our bodies and engaging one another and our technologies in less prescribed ways. That this may result in miscommunications is inevitable, but my anxiety is not the same as Norman's. Where he is looking to institute clear

28. Slavoj Zizek, How to Read Lacan (New York: W. W. Norton & Company, 2007).
29. From the opening statement of Mark Weiser's polemical article on ubiquitous computing, "The Computer for the 21st Century," Scientific American Volume 265, No. 3 (1991): 94-100. The entire statement reads, "The most profound technologies are those that disappear. They weave themselves into the fabric of everyday life until they are indistinguishable from it."

protocols for such systems, I am concerned with how the aesthetic potential of interacting ensembles will be compromised by unwarranted caution like his.

For the design of sentient cities that include responsive buildings, infrastructure, transportation and mobile devices, the concern for the moment should be on the kinds of interactions we want. This is a cultural question that requires us to probe the efficacy of interactive technologies and what it means for the ways in which we want to live. The role of designers, architects and artists in forecasting this is paramount, but caution needs to be taken in projecting uncritical utopias. The technologies are there — it is the imaginary of contemplating the interacting ensembles that is missing.

Whether NUIs, TUIs [30], GUIs or any other form of interfacing will be more adequate than the other will depend on the situation. Inevitably all will be in play, but to model the communication that we expect from such interactions on the workplace machine is misdirected. It will only force us into tempering interactivity and not recognizing its potential for displacing control and allowing unanticipated encounters to emerge. This potential is what designers, architects and artists aspire to in their work. It provides opportunities to delegate our agency to sentient systems, our activity as well as our passivity. Both yielding pleasurable and unpredictable results.

30. Abbreviation for Tangible User Interfaces. For more information see Hiroshi Ishii and Tangible Media Group, Tangible Bits: Towards Seamless Interface between People, Bits, and Atoms (Tokyo: NTT Publishing Co., Ltd., 2000).

NEW SPATIAL INTELLIGENCE, OR THE TREE ALLOWED TO GROW FREELY, BUT TO MAN'S PATTERN

DAN HILL

The Melbourne-based educator Leon van Schaik suggests architecture took a wrong turn when professionalizing in the mid-19th century, in thrall to the engineer of the emerging industrial economy. Van Schaik's critique is profoundly important, as it describes the seeds that have led to architecture's near-marginalization but also of its potentially influential future:

To compete with this practical glamour our forebears went to the heart of making in architecture — its technologies of carving, moulding, draping or assembling — when they staked their claim to be caretakers of a body of knowledge for society. The architectural capacity to think and design in three and four dimensions, our highly developed spatial intelligence, was overlooked, and for the profession space became, by default, something that resulted from what was construction... What if our forebears had professionalised architecture around spatial intelligence rather than the technologies of shelter? Might society find it easier to recognise what is unique about what our kind of thinking can offer?

The articulation and exploration of spatial histories that van Schaik suggests would be a fascinating next step for projects like those proposed in Toward the Sentient City. How might such projects help develop this understanding of spatial intelligence, and how it augments the kinetic, natural, linguistic, logical, mathematical, musical and personal intelligences? (In fact I'd hesitate before suggesting there is also an emerging 'informational' intelligence... but only just. Similarly, at other points a few of us have talked of understanding 'information as a material', and while this could conceivably lead to the digital equivalent of shelter-fixation, this too may be a theme worth developing.)

Developing a way of communicating such intelligences—and possibly related sensory modes such as an urban or informational form of proprioception — may be key to where these projects go, at least in hovering around an avant-garde that can generate useful prompts for the emerging mainstream business of urban informatics.

An axis that these projects positively explore, however, is that of seeing architecture itself as communication. I'm reminded of something the Swiss architect Bernard Tschumi said recently on the Melbourne radio show *The Architects*:

Architecture is a form of communication... of knowledge. Architecture is a way to understand our world, and also possibly to have some effect on it. It doesn't have to necessarily be through buildings — it has to do with ideas that involve our immediate environment, our physical space. Any way to use that physical environment, that architectural context, as a means to discuss issues I think is very appropriate.

That few of these projects concern a traditional understanding of the building, focusing instead on the implications of space and urban fabric 'becoming sentient', is thus hugely important. Tschumi's directive enables the profession to freely explore issues, ideas and interventions, only some of which need be built, or appear to be buildings. Designers could continue to productively plough this furrow, moving further beyond the traditional limits of building, articulating

1. Leon van Schaik, "Spatial Intelligence: New Futures for Architecture"

the implications of architects as conductors for spatial intelligence within a society and connecting ever deeper to an understanding of how people live in cities. These and other projects might then begin to match the leaps of imagination made by Peter Cook and Reyner Banham.

So let's look briefly at how these particular projects "use the physical environment as a means to discuss ideas," in Tschumi's words.

Amphibious Architecture is by The Living Architecture Lab at Columbia University (formerly The Living) and Natalie Jeremijenko of NYU and elsewhere. It's a rather beautiful piece of work, comprising two interactive networks of floating tubes, connecting the Bronx River and the East River. The tubes are both sensors and actuators, the latter in the form of LEDs, the former measuring water quality, presence of fish.

A hybrid development of earlier projects by both Jeremijenko and The Living, the glowing, bobbing lights are immediately compelling, but more interesting is this exploration of the relationship between urban bodies of water and the city. As Jeremijenko points out, there is a vast amount of public information about water quality yet this generates only small pockets of engagement. Meanwhile, waterfronts in New York and elsewhere have become attractive again to urban development, though only at a superficial level of the view over the water. "Harbour glimpses," as Sydney real estate agents say. This project moves beyond that facile surface connection to engage with water as a body, rather than a mirror with a memory. In doing so, it helps us understand something of what's going on in the opaque mass that created, nourished, supported and shaped New York.

The idea that you can "text-message the fish" is somewhat dubious, however. We've probably done enough to the poor buggers without now subjecting them to spam too. (Another Jeremijenko project has already suggested the fish in the Hudson River are on anti-depressants.) The project works perfectly well by remaining in the responsive, reflexive camp, without the need to suggest this faux-interactivity. There is something appealing about communicating directly with the sheer mass of river beneath the reflective surfaces, however, yet SMS as a medium doesn't quite seem up to it.

If SMS for fish is possibly a joke, then David Jimison's and JooYoun Paek's project *Too Smart City* certainly is. They deliberately use comedy to highlight the potentially absurd nature of placing too much hope in so-called smart technology, or rather outsourcing responsibility for civic spaces to algorithms and actuators. By subtly engaging with the political undercurrents in urban design, they've created a set of smart-arse street furniture that is all too plausible, embedding a sort of grumpy, obstreperous character into benches and trashcans, as if years of neglect had finally caused the city's street furniture to flip out.

The Smart Bench tips you up if you spend too long on it, disturbing potential vagrants. The Smart Sign berates you with legal codes. The Smart Trashcan spits your waste back at you if it's deposited in the wrong bin.

The artists describe the objects as "so smart that they're functionally useless," and it's good to see humor being deployed here. If Tschumi's directive to engage is taken seriously, then not being serious at all by going for laughs is entirely fair game, and probably essential, occasionally. Certainly these are close to the "funny because it's true" category. William H. Whyte's seminal 1970s photographs of New York are full of benches and ledges that are not supposed to be inhabited for any length of time — in fact you almost fear that this bench design may end up in the wrong hands and actually get commissioned. Similarly, the rhetoric around sorting one's waste can verge on a slightly desperate hectoring that makes the Smart Trash can seem like Oscar the Grouch in a rare good mood.

Though questioning what they describe as "the myth that technology is going to be a transformative force" is also fair game, it would be good to move beyond the question and at least hint at some answers. But as with many of the projects here, you can track ongoing progress on their entertaining blog and watch what develops, not least Jimison's attempts to create an "ass algorithm."

As if emerging from the rear end of the waste-bin in Too Smart City, the *Trash Track* is a continuation of the MIT SENSEable City Lab's research into sensor-based interventions in existing urban and telecommunications systems. In this case, trash is tagged with sensors and then followed throughout its slow, inexorable and often depressing journey to landfill or recycling. As with many projects in this area, it concerns the now-ubiquitous idea of making the invisible visible. In this case, the 'out of sight, out of mind' problem when it comes to dealing with waste. Some initial visualisations indicate an interesting product-centred view of how garbage interacts with the city, yet with this broader goal I'd be more interested in a sense of the aggregated footprint of the city itself, derived from all the coffee cups discarded over time.

The project suggests it's almost a practical demonstration of a core idea in pervasive computing — that of 'smart dust.' Though it isn't there yet — the sensors are large, cumbersome and apparently placed by hand — it hints at its possibilities, if remaining a little uncritical. The connection to architecture and urbanism is not as obvious, despite positioning garbage as an urban system, perceived in terms of mobility. I suspect the implications are more broadly for industrial design and product design. Again, though, it's worth reflecting on the sense of ambition, scale and connectivity that often comes with MIT's work. Their ability to connect does place them in a sphere of influence, and at the highest levels.

Usman Haque, a designer who in the creation of the Pachube data-meets-place-meets-people platform has grasped the promise of informatics as much as anyone, makes a typically fascinating contribution. *Natural Fuse* is simply more interactive — it's not just an abstract signal that might raise awareness around energy consumption, and so possibly stimulate intervention, but actually and actively requires an intervention of the users. The possible outcome of the *Natural Fuse* system — that PLANTS WILL DIE! if energy consumption gets too high — may be the more visceral mode of engagement that is required around energy consumption, as compared to the quietly glowing LED screens of smart meters that have replaced the flashing '12:00' of VCRs in homes around the land. (Though you half-wonder whether an 'arms race' of ever more gruesome natural deaths may be required if people become inured to wilting brown husks of Leopard Lily?!)

The project also describes an architecture around the Toward the Sentient City body of work — of websites, RSS and Twitter feeds, maps and of the platform of Pachube itself. Again, the team led by Haque appear to have the greatest facility with this networked aspect of informatics. And this is interactive, at least to some degree, where others are responsive. Whether there is a spatial intelligence being articulated here is another matter — this is perhaps more informational intelligence than spatial. And so is it architecture? Many of Haque's other works certainly are, rather more obviously, but his sense of "the software of space" is certainly interesting and represents a genuine attempt to reconfigure architecture around an understanding of the shared and ongoing production of cities. Watch that (soft) space.

The most genuinely interactive contribution of all — in that it comprises an essentially open and accessible platform — is from the team led by Anthony Townsend and featuring several from workplace consultants DEGW. *Breakout!* is inspired by the co-working spaces (or "jellies" as some are known) that enable work to take place via shared resources. It's a mobile infrastructure of power, connectivity and social interaction, supported by smart social software application to facilitate transient workplaces. While the stated motive is to "liberate office workers from office buildings," it could be argued that this is already happening — see all the ventures the Breakout! team is inspired by, for a start. When it doesn't happen, it's not for lack of infrastructure in a city like New York, but rather social, cultural and economic issues around present-day business culture. However, the Breakout! team can hardly change that overnight, and like the annual PARKing Days, which temporarily reappropriates on-street parking for a variety of playful and productive uses, we're reminded of the value in creating a series of public focal points on such activity, and so of the powerful role of events in shaping the city.

In terms of the environments the Breakout! team can create, I'd like to see a more varied set emerging, with different kinds of workshops available, particularly those that involve physical 'making' via the return of light manufacturing to urban environments. But that can come later.

Almost all of these commissions do variously deal with the social rather than built form, with explorations of the civic realm rather than physical structure. Sure, there are a few shortcomings but taken as a whole they comprise an excellent set of examples of how a more sophisticated relationship between architecture and the sentient city might unfold over the next few years.

"Might" is a key word there, for it's going to take a lot of effort to retrain and reorient architects in general for this particular form of spatial intelligence, and while works like these are part of that effort, they're not enough in themselves.

The entire practice of architecture and urban design, its sensibility and economic model may need redressing (as with many other fields, of course.) Given their previous predilections, the lack of technical and conceptual understanding — never mind an apparently congenital inability to design a decent website — means the profession has a long way to go before it can demand a seat at the table. An admittedly fading tradition of thinking of itself as the 'master builder' needs to be entirely excoriated once and for all. Devising the architect's new sensibility — what Paul Dourish would describe as "the designer's stance" for the discipline — will also be fundamentally important. Either way, complex urban systems are well beyond the ken of the sole master builder; they have been for years, but increasingly so with this ever more multi-layered understanding of the city.

Other design disciplines — interaction design, industrial design, service design, to name three — are currently far better placed to lead on these ideas, within multidisciplinary design teams. So the architect may be best-placed as part of that team, leading on spatial intelligence just as others might lead on information and communication systems, materials, structures, embodied interaction, behavioral psychology, topography, acoustics, biodiversity and so on. In a recent conversation with the SENSEable City Lab's Carlo Ratti, we ended up sketching out a loosely multidisciplinary team in which the architect was one of perhaps ten different disciplines, all of whom would lead at various points.

Yet there doesn't seem to be much explicit recognition of that in these projects. I'd like to have seen more of a debate of the craft, process and shifting nature of disciplines throughout. And of course the smart, progressive and more inclusive designers represented here are far-removed from the Roark stars of architectural mythology. But whether many in the wider profession have moved far enough in their direction is another matter.

To put it another way, who would you *actually* have design some sentient street furniture? Naota Fukasawa or Frank Gehry? Jonathan Ive or Daniel Libeskind? Luigi Colani or Ken Yeang? (Actually, I suspect that the intentions of any sentient furniture designed by Colani may not be entirely honorable. Wipe-off plastic required. Be careful where you sit.)

But given that it's actually the likes of large software and hardware corporations that are currently attempting to claim the 'smarter cities' mantra for their own, I'd like the profession to pull up a chair nonetheless. At least many architects and urban designers have an understanding of what makes good cities tick.

With that in mind, these projects are fantastically useful and worthwhile. They weave together several threads running through contemporary thinking around cities and information, and through the provision of context and discussion, they do indeed use architecture, of a sort, as a means to discuss ideas, as Tschumi suggests, or to more broadly exert a form of spatial intelligence, after van Schaik. In this, they suggest the new form of architecture that we're all striving for, something beyond variations on shelter or indeed the gaudy showmanship of the last two decades, in which new ways of living can be articulated via a genuinely open and interactive framework.

Devising an urbanism that extends this last aspect — urban fabric suffused with rich forms of civic interactivity — is perhaps the biggest challenge. Funnily enough I'm reminded of the British architecture critic Ian Nairn's line about the avenues of trees in Bushey Park, surrounding Hampton Court — and by extension England, and design:

Man proposes, and a noble enough proposal. Nature takes over, but heeds man's direction. All you see now are the glorious trees: but they would not have been so glorious without the initial design. As a symbol of Hampton Court, and of the whole of England, you could do worse: the tree allowed to grow freely, but to man's pattern.

Perhaps in this line from 1966, concerning a design from the 1530s, there are the seeds of a more responsive and sometimes interactive architecture, iterative in development, predicated with change in mind, a space described around both biological and social ecosystems, constructed around the organic, chaotic and complex, yet symbiotically shaped by clarity and intent.

Van Schaik implores us not to unthinkingly overlay inappropriate spatial histories and asks instead "What spatial constructs does this new place allow?" We should ask the same of such rare opportunities for addressing the promise of the sentient city. These stimulating works begin to tentatively trace out some lines describing this new place, and those of us who inhabit cities will benefit hugely from continuing to pick apart their implications.

BOXES TOWARDS BANANAS: SPATIAL DISPERSAL, INTELLIGENT CITIES AND ANIMAL STRUCTURES

MATTHEW FULLER

One of the genesis stories of architecture describes a first moment of shelter around a fire whose focus forms a locus of gathering within the radius of its warmth. Humans or Neanderthals congregate around a hearth, drape furs across overhanging branches or recede into caves and eventually find themselves building huts, houses, palaces and office blocks in order to maintain the warmth of the blaze. Reyner Banham starts his examination of the thermal qualities of late modern buildings with just such a tale.[1] Architecture is, in this sense, about maintaining warmth profiles and 'thermal delight.'[2]

In contemporary architecture, the question of heat and of flows within space has not only been maintained, but has come into play with full force in relation to the problems of climate damage. Whereas it can be used as a means of driving an aesthetics of austerity, it is also an aspect of building which emphasizes the sensual and physiological as deeply imbricated in the technical; the simple or complex means by which air forms part of the stuff of building, generating the possibilities for new forms and relations.[3] But Banham's tale is also about architecture as a means of generating foci, centers, gathering places, spaces of the coming together of energies, social arrangements, resources. If architecture has the act of coming together around a fire at its vestigial core, it is also entangled with the intricate iterations of social relations that are implied by any decision or need to gather and the kinds of conviviality, power and intelligence that are implied by them.

In 'Understanding Media,' Marshall McLuhan develops this theme, but shows how both media systems and building technologies operate to spread things out as well as to concentrate them.[4] The family spreads out from the hearth thanks to central heating, but is brought back together to gaze at the flickering embers of civilization on the television, before separating again to catch up with their networks via cellular. Architecture is also then about dispersion, separation, the sorting out of things in relation to space. This becomes especially the case when architecture and media systems start to mutually influence each other. Indeed, communications, a term that originally included both media and transportation, did (in the work of nineteenth century technocratic visionaries such as Saint-Simon amongst others) establish a mutual dynamic of differentiation and unification as crucial to the understanding of these systems.[5] The mutually intensifying relation between the telegraph and the train is a well-known example. McLuhan inevitably links this to the question of culture, and the kinds of intelligence and understanding each of these compositions fosters. The sprawl-cycles of North American cities, involving cars, shopping malls, television, freeways, advertising, the development patterns of suburbs and exurbs, telephones, and families going nuclear exemplify another complex set of patternings between spatialities and media systems.[6] The debate about the implications for intelligence given such examples tends to fall either towards the mournful or the chiliastic, but perhaps there are other directionss to take.

If we also attend to the dynamics of dispersion as a formational impulse in architecture and reflect on the implications

1. Reyner Banham, Architecture of the Well-tempered Environment, Second, Revised Edition. Architectural Press, London, 1984.
2. Lisa Heschong, Thermal Delight in Architecture, The MIT Press, Cambridge, 1979. For an analysis of bee construction in relation to its thermal qualities, see, Mathis, R. C. and D. R. Tarpy. (2007). "70 million years of building thermal envelope experience: building science lessons from the honey bee", ASHRAE Journal, no.49, pp1-8
3. Architects working on such themes include, Michelle Addington, Steven Gage, Sean Lally, (www.w-e-a-t-h-e-r-s.com/) and Philippe Rahm, http://www.philipperahm.com/
4. Marshall McLuhan, Understanding Media, the extensions of man, The MIT Press, Cambridge, 1994
5. Armand Mattelart, The Invention of Communication, trans. Susan Emanuel, University of Minnesota Press, Minneapolis, 1996
6. Margaret Morse, 'The ontology of everyday distraction: the freeway, the mall and television', in Patricia Mellencamp, ed., Logics of Television: Essays in Cultural Criticism, Indiana University Press, Bloomington, 1998, pp193-221

it might have for understanding media ecologies, information and cognition in urban contexts, what happens? A great deal of middens, burial grounds, sacred zones, gardens, and other kinds of specialized sorts of space generated as humans became sedentary also become key to the genesis of architecture. Rather than one of gathering, this is an architecture of things to be kept apart.

The media philosopher Vilém Flusser suggests that the development of spaces requiring separation is closely linked to the genesis of technology. In a text on the nature of 'factories', such as sites where flint tools or clay pots might be worked, and which in turn introduce new kinds of organization, he suggests that, "As soon as tools are introduced, specialized factory areas can and must be cut out of the environment."[7] What are now archaeological sites rich with flint fragments or pot sherds are the kinds of factories he has in mind. One of the dynamics leading into the generation of architectural morphologies is that generative of finer and finer grains of spatial differentiation. Because tools amplify and mutate human capacities, they require special places for the kinds of wastes they produce, the strange and dangerous movements they encourage and the kinds of resource aggregation that they require.

So if we are to follow the argument proposed by the archaeologist Leroi Gourhan, that the evolution of tools is both the prompt for and the symptom of the evolution of intelligence, the creation of spaces for the working of tools is part of the wider set of physical consequences of such changes.[8] If tools are things that denaturalize the human, they are perhaps also things that draw them away from the family and into new artificial kinds of space. Here, the evolution of intelligence also potentially becomes a matter of spatialization.

If we are to momentarily move away from the scale of the human however, and to imagine other genesis stories for architecture, those that arise from the point of view of other species, what might result, and what kinds of patterns of gathering and dispersal would appear? Might they suggest potentially novel kinds of architecture that are of interest in relation to the sentient city? Doing so implies a biomimesis, not solely at the level of morphology, but also at that of organization. Such an approach has a history of course, most conspicuous in politics [9] and more recently in robotics.[10] Here we are also able to draw on the building work of other species, in the ways that, according to kind, they may gather and sort waste, dry bedding, spatialize the rearing of young, communicate and in many different ways, shape spaces according to function whilst themselves in turn providing housing for parasites and pathogens. What forms of space are imaginable with a non-human genesis story which also attends to selection, sorting and spatial distribution as a variant root of architecture? Further, given an architecture founded on such a spatial variation, what kind of intellection and information systems might be entailed?

It is well known that some of the most enormous worked structures on the surface of the planet are those made by non-human species. Of these, giant beaver dams and coral reefs are visible to satellites. Smaller constructions, including termite

7. Vilém Flusser, The Shape of Things, trans. Anthony Mathews, Reaktion Books, London, 1999, p46
8. André Leroi-Gourhan, Evolution et Techniques. L'Homme et la matière, Albin Michel, Paris, 1971
9. Bernard Mandeville, The Fable of the Bees, or Private Vices, Publick Benefits, Penguin, London, 2007; Yann Moulier Boutang. "L'Irruption de l'écologie ou le grand chiasme de l'économie politique", in, Multitudes 24: Spring 2006. The specific integration of bees, the insect that provides itself for so many as a metaphor for economics, into capitalism as a commercial product has distinct and catastrophic consequences as colony collapse disorder, where the over-working, undernourishment and overconcentration of bees renders them subject to potentially exterminating disease.
10. For a survey of such work see John Johnston, The Allure of Machinic Life, cybernetics, artificial life and the new AI, The MIT Press, Cambridge, 2008. Particularly relevant is chapter seven's discussion of behaviour-based robotics.

hills of immense structural and functional sophistication, and the temporary architectural mating displays of bower birds, are prominent examples of non-human building.[11] The architectural fecundity and brilliance of non-human species is diverse and woven over time into human understanding of settlement and media.[12] One form of dispersal amongst animals is in the treatment of dead. W. G. Sebald draws on this in his book of meditative walks through the English county of Suffolk a region that has, over the last half-millennium, been incrementally abandoned. Here, amongst the empty fields, Sebald muses on the Baroque divagatory work *Hydriotaphia* or *Urn Burial* which makes reference to, "sepulture in elephants, cranes, the sepulchural cells of pismires and practice of bees" in its survey of death rituals.[13] Whilst poetically resonant, elephants' graveyards are now regarded as being mythical, the gathering of several skeletons being more likely the result of drought and suchlike. But the habits of ants, (pismires) who separate and gather the corpses of their dead at some distance from the nest bear out such musings. The differentiation of the dead from the living is in many species an inherently spatial activity, in some perhaps crucially so.[14]

Bees, ants and other social insects are often used as a means of describing kinds of non-linear collective decision-making, both through their navigation-oriented language and the pheremonal communications they generate in collectively modulating the construction of hives or nests. This organizational thematic of emergence is seen as characterizing behavior in which multiple scales of decision making, communication, and transformation, placing and modification of materials, occurs through the interactions of the bees, without any central force of command, to achieve sophisticated and critical tasks.[15] In the case of bees, one of these tasks is the splitting of hives, an act of dispersal that, through swarming, founds a new hive in response to overcrowding or other factors. As an event, it is one of risk, in that a newly built hive entails possibilities of failure greater than those of an established one, but splitting up is essential to the prosperity of the species and for the health of a hive that is growing beyond its resources.

The architecture of animals that move, that spend the year in a cycle of migration is often very different from that of those engaged to a territory. For such species, cocoons, nests, eggs and holes bored into cliffs, are temporary structures, of greater or lesser intimacy to the body. The more an animal moves around the surface of the earth, the less its need or capacity to stabilize a niche, instead it moves to more favorable climes or spaces. In this sense, constant movement is an architectural strategy: think of the swift which flies around the globe, but which returns annually to the same nesting site.[16]

Other forms of dispersal in relation to structure can be highly sophisticated and of necessity differ from species to species. A quick survey of a range of nest structures can show this more usefully than any single, and thus 'exceptional' example. The Sociable Weaver for instance is an African bird that collectively builds enormous multi-chambered nests sometimes almost as big as the tree they are built into. These nests act as

11. These are amongst those catalogued in, James R. Gould & Carole Grant Gould, Animal Architects, Building and the Evolution of Intelligence, Basic Books, New York, 2007. See also, Karl von Frisch, Animal Architecture, Helen and Kurt Wolff, New York, 1974

12. Bees, the insect that probably yields the most metaphors in human cultures have, as Juan Antonio Ramirez notes, historically been one of the most significant species for the understanding of architecture and organisational form. (See, Juan Antonio Ramirez, The Beehive Metaphor: from Gaudi to Le Corbusier, trans. Alexander R. Tulloch, Reaktion, London, 2000) In media theory, they, and other insects, are the subject of sustained archaeological speculation by Jussi Parikka, yielding insights into the history and imaginary of media. (Jussi Parikka, Insect Media: An Archaeology of Animals and Technology, University of Minnesota Press, Minneapolis, 2010). Both insects and pigeons figure amongst the catalogue of submerged influences on architecture brought together by David Gissen. (David Gissen, Subnature, architecture's other environments, Princeton Architectural Press, New York, 2009)

13. Winfried Georg Sebald, Rings of Saturn, trans. Michael Hulse, Vintage, London, 2002, p.25

14. The legendary dog that, with what might be called a pre-emptive anthropomorphism, mourns its human at graveside is an intriguing counter-case to death requiring spatial separation. See, i.e. the story of 'Greyfriars Bobby' in E.S. Turner, All Heaven in a Rage, Centaur Press, Fontwell, Sussex, 1992

15. Nigel R. Franks & Jean-Louis Deneubourg, "Self-organizing nest construction in ants: individual worker behaviour and the nest's dynamics", Animal Behaviour, 1997, vol.54, pp.779–796

16. See, i.e. London Swifts, an organization advising on the maintenance of building space for these birds. http://www.londons-swifts.org.uk/

heat regulators so well that the body-heat of the birds is sufficient to warm them overnight. In summer it is usual that only two birds may share a chamber, whilst in winter, five will roost together.[17]

Dispersal may also take place as a way of evading predators. Occasionally this imperative will have architectural consequences. The male of the Orange Cheeked Waxbill constructs small decoy nests filled with odorous dung in order to draw snakes away from its young. The Baya Weaver produces highly shaped pendular nests that hang from branches, often containing more than one chamber, leaving time for escape if one is invaded. [18] The decoy dwelling is an unusual, if not entirely unknown, category in human architecture more often belonging to the genre of fortification,[19] but it does prompt the question of correlations to existing or new typologies of use and organization in media and architecture, and suggest another means by which a kind of intelligence is spatiality manifest. What is different in the medial components and articulation of animal architecture? One quality is that of bodily intimacy, we tend not to chew and excrete our building materials, as some birds or insects do, partly because, having hands, we have other means of carrying and placing materials. Nor, as in the case of beavers who use their dams and lodges as effective foodstores over winter, do humans often use their own buildings as food, the popular images of folktales such as Hansel and Gretel, full of the enticing evil of houses made of gingerbread and icing, notwithstanding.

Moving a little further from the sphere of the built structure as an artifact to an understanding of it as media, numerous types of spider maintain relations of sensing and calculation through the webs they produce. These delicate and powerful constructions exist both as means of capture and of remote sensing, with spiders able to calculate the position of prey, its size, and relative degree of entrapment and liveliness. The spider maintains a position to one side of the web, its feet touching an array of strands to check their displacement. In that "a subtle portrait of the fly"[20] is drawn in the web of the spider, this is also a system that evinces proper medial qualities of integration and communication, whilst at the same time promising the dissolution of the domains previously internal to that which is drawn into communication. Sensual extension, capture and the precise delineation of space in a spontaneous, tirelessly reworked and cunningly arranged net is crucial to the medial trope of dispersal. Become removed, the better to be able to entangle your prey.

If we turn now to how media systems may be seen to generate their own kinds of spatiality, a key trope in recent decades has been the mutual describability of media, information and space in terms of flow, the reimagination of urban and medial space as a kind of generalized cornucopian conduit. Here, dispersal tends towards existence under the conditions of universalization, or in turn, of a condition of super-connectivity, lack of access to which entails structural deprivation. In terms of re-imagining the arrangements of computation and sentience in the city therefore, such an approach suggests a supplementary

17. Gordon Lindsay Maclean, The Ecophysiology of Desert Birds, Springer, Vienna, 1996
18. The extent to which the architecture of other species is inherited or learned remains the subject of debate. The modern form of this debate has a beginning with the early investigations into animal intelligence. For an articulation of this discussion in relation to nest building see, i.e. Patrick T. Walsh, Mike Hansell, Wendy D. Borello and Susan D. Healy, 'Repeatability of nest morphology in African weaver birds', Royal Society Biology Letters, 2009.0664. The Linnean names of the species mentioned are: the Swift, family *Apodidae*; Sociable Weaver, *philetarius socius*; Orange Cheeked Waxbill, *estrilda melpoda*; Baya Weaver, *ploceus philippinus*.
19. See for examples of such in recent protest constructions, Nils Norman, The Contemporary Picturesque, Bookworks, London, 2000
20. Gilles Deleuze and Félix Guattari, What is Philosophy?, trans. Graham Burchell and Hugh Tomlinson, Verso, London, 1994, p.185

Baya Weaver nests. Photo Bharat Patel.

dynamic, one of clustering and layering around information sources. Whereas accounts of locative media often imagine geographically defined data sources as encouraging such clustering, to make space more homely, to tag it and anchor it in relation to grids, in order to understand such media in relation to dispersal, it is useful to understand such spatializations as themselves having a certain kind of capacity to spread. Such spread however is also one of deeply uneven distribution, such imbalance or inequity providing the motive forces for other kinds of abstract development, such as those of financial yields.[21]

The relative capacity of certain forms of space to migrate and to reformat other zones can also be found in the dense micropolitics of encoded spaces where information architectures combine with urbanism in such a way that concentrations, separations and chains of concatenating functions allow for uncanny connections to spread themselves across space and to produce spaces. Timetables, databases and maps all generate their own kinds of spatialization and arrangement of information, relations between columns and rows, a spatiality of information, that is then in turn reinvested in other forms of space including the urban. The newer electromagnetic geographies of permissions, locations, checking and timing, of modulations, frequencies and differential access, acceleration and accentuation, in turn feeding into screen-size landscapes of pixels, vectors and cells with variably linked functions, arrays and outputs, provide a means for understanding both the way things are spread out in space, and how different modes of spatiality themselves spread, concentrate and decline.

But what can we draw from this quick sprint away from the hearth and meander through animals' productions of space? An initial comment is that space is, in certain ways and to differing degrees, species-specific. Each landscape reveals affordances and dangers that, like the web to the fly, are significant only to certain sensorial natures, intelligences and capacities. Simultaneously, all spaces are generated out of the layered and multi-scalar interactions and indifferences of a wider ecology.

21. David Harvey, Spaces of Global Capitalism, towards a theory of uneven geographical development, Verso, London, 2006

The tangled natures of ecologies are also those in which various kinds of forces, intelligences, drives and physiologies try to figure each other out, each in turn conditioned by their fitness to environment and their incipient, overt or hidden capacities to change that environment. Interactions between the dynamics of concentration and dispersal, between gathering and flight, appearance and hiding provide a characteristic element to such tangled landscapes which are always more than this brief listing of tendencies can suggest.

For animals, the understanding of space is intimately connected with the ways in which it is inhabited, made, experienced. This is also true for humans, but the degree of complication attendant to all of these is exponentially amplified in relation to synthetic systems such as the architecture and media specific to them. Any one of these systems is beyond the state of knowledge of any individual, and as such they inherently require collaboration, or mechanisms of obfuscation. In such a context however, human intelligence and manipulation of tools is no longer selected by fitness to environment but by more or less tenuous gambles on the possibility of fit with, avoidance of, or triumph over existing systems and their ability to mobilize, extract or inspire certain other kinds of coefficiency or scalar autonomy. This perpetually displaced and uneasy intelligence in turn generates demands for momentary, perhaps market-driven, intellectual fitnesses that are soon rendered redundant, rotten or exemplary. Attention to the multiple compositional dynamics of intelligence in forms of spatiality may in the present day reveal a new typology by which humans find themselves tested for fitness: rather than the peppy irritation of smartness, a preponderance of stupid spaces.

Interestingly, the debates around the extent and nature of animal intelligence that started in scientific earnest at the beginning of the twentieth century took a decidedly architectural form. Early animal psychologists, such as Edward Thorndike (also notable as a founder of psychometrics) made use of contraptions such as mazes, where cats or rats were left to find their own way out, or cages and boxes with walls or panels which, when pressed, revealed food or exits.[22] Intelligence, understood to be a capacity for learning, was measured by the decrease in the amount of time that each animal took to activate the specified mechanism. Measured and transposed as data to another form of spatial mechanism, the graph, this behavior introduced the phrase 'the learning curve' into popular vocabulary. The model of animal intelligence offered by these researchers was countered by others, such as the Gestalt Psychologist Wolfgang Köhler.[23]

Stranded along with a group of chimpanzees on Mallorca during the First World War, Köhler set up a series of experiments in which the chimps were provided with basic construction materials, such as stackable crates and bamboo poles, and a problem — bananas suspended high above them on a rope. In what remains a set of famous experiments, the chimpanzees, after forethought and testing, eventually developed a way to make a tower of two, three, or four crates, and knock the bananas down with a pole. Köhler claimed that it was highly

22. Edward Thorndike, Animal Intelligence, Macmillan, New York, 1911
23. Wolfgang Köhler, The Mentality of Apes, Harcourt Brace, New York, 1925

PLATE VI. CHICA BEATING DOWN HER OBJECTIVE WITH A POLE

Wolfgang Köhler's basic construction experiments with chimpanzees. From The Mentality of Apes, W. Koehler, Copyright 1956, Routledge & Kegan Paul. Reproduced by permission of Taylor & Francis Books UK.

PLATE IV. GRANDE ON AN INSECURE CONSTRUCTION (NOTE SULTAN'S SYMPATHETIC LEFT HAND)

Wolfgang Köhler's basic construction experiments with chimpanzees. From The Mentality of Apes, W. Koehler, Copyright 1956, Routledge & Kegan Paul. Reproduced by permission of Taylor & Francis Books UK.

unlikely that such a result could be achieved by chance. He proposed instead that the primates impelled themselves to think through the numerous possibilities of the situation in order to gain "the objective." Here, a dynamic space is generated by the ability of the chimpanzees to handle materials, to work with the compositional possibilities of crates and poles, to imaginatively interpolate themselves, and to be mobilized by their lust for bananas.

The argument between these two approaches is usually staged as between trial and error versus intuition.[24] But perhaps we can also see this debate in early modern psychology as embodying an architectural problem, one that is deeply involved in the kinds of intelligence our cities and spaces require, generate, and involve. Whether we make mazes and trick boxes, or structures that require fundamental thought and a process of enquiry and redesign, implies an understanding of space as inherently involved in thought processes. The quest for bananas stacks up questions about the kinds of thinking that cities are able to make space for.

Learning from such non-human spatialities, how is it possible to imagine and to develop architectures, urban designs and modes of thought about cities that take part in realizing the question of intelligence as implicated and involved in differing spatial dynamics?

Firstly, one of the most urgent means of developing such an approach is by engendering a sensitivity to the urban in which multiple kinds of intelligence, including those of non-human species and their spatial practices in all their fundamental alienness to humans, have a significant place rather than being treated as simply pestilent, indifferent or decorative.

Secondly, to recognize that in the generative development of spatialities that intensify intelligence, specialization takes place, (as in the case of places for the manufacture of tools). Cities can be characterized as a concentrated process of the gathering, enfolding and dispersal of such spaces. In becoming strange themselves through such specialization and congruence, they create mutant fitness landscapes for forms of intelligence to interpret, cohabit, or to disperse from.

Thirdly, how can we elicit the genesis of a city that in its morphological aspects is also a city that 'learns', that engages in relations to space and the developments of spatialities that are capable of adaptation and experiment at multiple scales and that thus also incite new learning? Part of the work of asking such a question is in recognizing a vital conceptuality of life that is also spatial, and that spatialities have their own kinds of vitality, capacities of repetition, variation and adaption, that in turn feed back into the becomings of sentience.

24. A contemporary portrait of this debate can be found in Bertrand Russell, An Outline of Philosophy, George Allen and Unwin, London, 1927

UNSETTLING TOPOGRAPHIC REPRESEN-TATION

SASKIA SASSEN

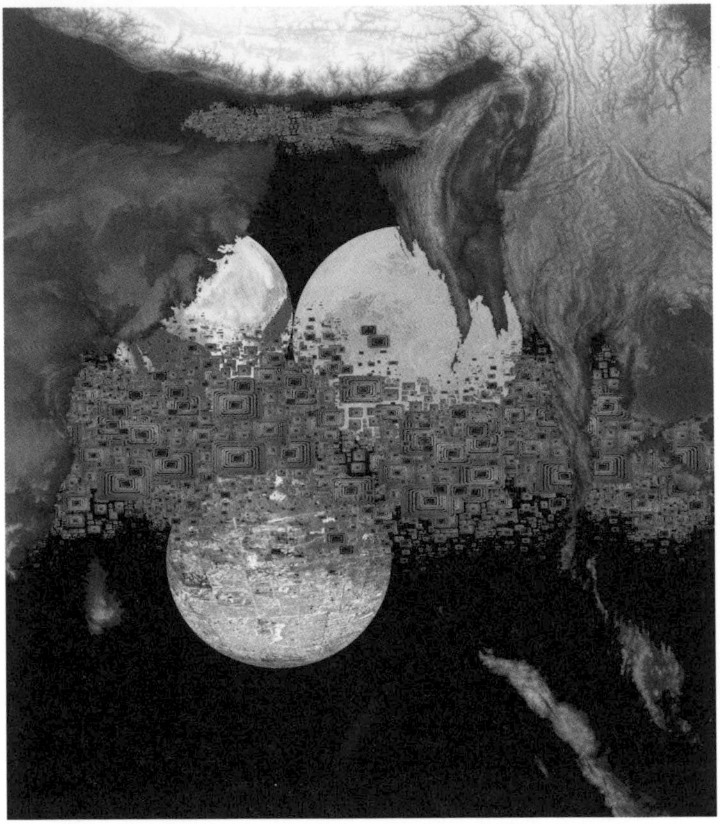

Hilary Koob-Sassen (www.TheErrorists.com) presented in "When Territory and Time Seep out of the Old Cages," Time Marathon, Solomon R. Guggenheim Museum, Jan 6, 2009.

The topographic moment is a critical and a large component of the representation of cities. But it does not easily incorporate the fact of globalization and digitization as part of representing the urban. Nor can the topographic give us the tools to contest today's dominant accounts about globalization and digitization which evict place and materiality. This ineffectiveness holds *even* in those instances when globalization and digitization are deeply imbricated with the material and the local and hence with that topographic moment.

A key analytic move that bridges between these very diverse dimensions is to capture the possibility that particular components of a city's topography might be spatializations of global and digital dynamics and formations. Such particular topographic components would then be one site in a multi-sited circuit or network that also includes non-topographic and non-urban components — topography as one element in a larger assemblage of material and digital components. Cities today are increasingly such assemblages.

Spatialized Power Projects

Cities have long been key sites for the spatialization of power projects-whether political, religious, or economic. There are multiple instances that capture this spatialization in cities and metropolitan regions. We can find it in the structures and infrastructures for the control and management of past colonial empires and of current global firms and markets; we can also find it in the segregation of population groups that can consequently be more easily produced as either cheap labor or surplus people; in the choice of particular built forms used for the representing and symbolic cleansing of economic power, as in the preference for "Greek temples" to house stock markets; in what we refer to today as high-income residential and commercial gentrification to accommodate the expanding elite professional classes, with the inevitable displacement of lower-income households and firms. We can also see it in the large-scale destruction of natural environments to implant particular forms of urbanization marked by spread rather than density and linked to specific real-estate development interests, such as the early uncontrolled strip-development and suburbanization in the Los Angeles region.

Yet, the particular dynamics and capacities captured by the terms globalization and digitalization signal the possibility of a major transformation in this dynamic process of spatialization. The dominant interpretation posits that digitalization entails an absolute disembedding from the material world. Key concepts in the dominant account about the global economy-globalization, information economy, and telematics — all suggest that place no longer matters. And they suggest that the type of place represented by major cities may have become obsolete from the perspective of the economy, particularly for the leading industries, as these have the best access to, and are the most advanced users of, telematics.

These are accounts that privilege the fact of instantaneous global transmission over the concentrations of built infrastructure that make transmission possible; that privilege information outputs over the work of producing those outputs, from specialists to secretaries; and that privilege the new transnational corporate culture over the multiplicity of cultural environments, including reterritorialized immigrant cultures, within which many of the "other" jobs of the global information economy take place. [1]

One consequence of such a representation of the global information economy as placeless would be that there is no longer a spatialization of this type of power today: it has supposedly dispersed geographically and gone partly digital. It is this proposition that I have contested in much of my work, arguing that this dispersal is only part of the story and that we see in fact new types of spatializations of power. [2]

Unstable Meanings

Such spatializations of the global and the digital destabilize the meaning of the local or the sited, and thereby of the topographic understanding of cities. It is probably especially in global cities that these spatializations are common. My concern here is to distinguish between the topographic representation of key aspects of the city and an interpretation of these same aspects in terms of spatialized global economic, political, and cultural dynamics. [3] This is one analytic path into questions about cities in a global digital age. It brings a particular type of twist to the discussion on urban topography and cities since

1. This privileging entails the eviction of a whole array of activities and types of workers from the account about the process of globalization, which, I argue, are as much a part of it as is international finance. The eviction of these activities and workers from the dominant representation of the global information economy has the effect of excluding the variety of cultural contexts within which they exist, a cultural diversity that is as much a presence in processes of globalization as is the new international corporate culture.

2. See Saskia Sassen, The Global City, rev. ed.(Princeton: Princeton University Press, 2001).

3. These are all complex and multifaceted subjects. It is impossible to do full justice to them or to the literatures they have engendered. I have elaborated on both the subjects and the literatures elsewhere (Sassen 2009: chapters 7 and 8). Other important developments for understanding the city which I will not discuss here but have developed elsewhere are the urban consequences of several current major crises, such as asymmetric war, which makes the city a technology for war (2010), and the financial crisis that takes off in 2008 (2009a).

globalization and digitization are both associated with dispersal and mobility.

What analytic elements need to be developed in order to compensate for or remedy the ineffectiveness of topographic representations to make legible the fact that at least some global and digital components get spatialized in cities. Among such components are prominently the distributed presence of digitized capabilities. But also more hybrid, and often elusive capacities, such as the power projects of major global economic actors as well as the political projects of contestatory actors, e.g. electronic activists. A topographic representation of rich and poor areas of a city would simply capture the physical conditions of each — advantage and disadvantage. It would fail to capture the electronic connectivity that can make even poor areas into locations on global circuits. Once this spatialization of various global and digital components is made legible, the richness of topographic analysis can add to our understanding of this process. The challenge is to locate and specify the fact of such spatializations and their variability.

This brings up a second set of issues: topographic representations of the built environment of cities tend to emphasize the distinctiveness of the various socio-economic sectors: the differences between poor and rich neighborhoods, between commercial and manufacturing districts, and so on. While valid, this type of representation of a city becomes particularly partial when , as is happening today, a growing share of advanced economic sectors also employ significant numbers of very low-wage workers and subcontract out to firms that do not look like they belong in the advanced corporate sector; similarly, the growth of high income professional households has generated a whole new demand for low-wage household workers, connecting expensive residential areas with poorer ones, and placing these professional households on global care-chains that bring-in many of the cleaners, nannies and nurses from poorer countries. In brief, economic restructuring is producing multiple interconnections among parts of the city that topographically look like they may have little to do with each other. Given some of the socio-economic, technical, and cultural dynamics of the current era, topographic representations may well be more partial today than in past phases.

The limitations of topographic representations of the city to capture these types of interconnections —between the global and the urban, and between socio-economic areas of a city that appear as completely unrelated and the world — call for analytic tools that allow us to incorporate such interconnections in spatial representations of cities. Some of these interconnections have long existed. What is different today is their multiplication, their intensity, their character. Some elements of topographic representation, such as transport systems and water and sewage pipes, have long captured particular interconnections. What is different today in this regard is the sharpening of non-physical interconnections, such as social and digital interconnections. Perhaps this also points to a deeper transformation in the larger social, economic and physical orders.

Topographic representations remain critical, but are increasingly insufficient. One way of addressing these conditions is to uncover the interconnections between urban forms and urban fragments, and between orders — the global and the urban, the digital and the urban—that appear as unconnected. This is one more step for understanding what our large cities are about today and in the near future, and what constitutes their complexity.

Analytic Borderlands

As a political economist, addressing these issues has meant working in several systems of representation and constructing spaces of intersection. There are analytic moments when two systems of representation intersect. Such analytic moments are easily experienced as spaces of silence, or of absence. One challenge is to see what happens in those spaces, what operations take place there. In my own work I have had to deal frequently with these spaces of intersection and conceive of them as analytic borderlands —an analytic terrain where discontinuities are constitutive rather than reduced to a dividing line. Thus much of my work on economic globalization and cities has focused on these discontinuities and has sought to reconstitute their articulation analytically as borderlands rather than as dividing lines. [4]

Methodologically, the construction of these analytic borderlands pivots on what I call circuits for the distribution and installation of operations; I focus on circuits that cut across what are generally seen as two or more discontinuous "systems," institutional orders, or dynamics. Instead I try to capture the distributed format of so much of what we think of as a point in space. These circuits may be internal to a city's economy or, perhaps at the other extreme, global. In the latter case, a given city is but one site on a circuit that may contain a few or many other such cities. And the operations that get distributed through these circuits can range widely— they can be economic, political, cultural, subjective.

Circuits internal to a city allow us to follow economic activities into territories that lie outside the increasingly narrow borders of mainstream representations of the urban economy and to negotiate the crossing of discontinuous spaces. For instance, it allows us to locate various components of the informal economy (whether in New York or Paris or Mumbai) on circuits that connect it to what are considered advanced industries, such as finance, design or fashion. A topographic representation would capture the enormous discontinuity between the places and built environments of the informal economy and those of the financial or design district in a city, but would fail to capture their complex economic interactions and dependencies.

International and transnational circuits allow us to detect the particular networks that connect specific activities in one city with specific activities in cities in other countries. For instance, if one focuses on gold futures markets, cities such as London and Frankfurt dominate; and if we focus on the metal, then these are joined by Sao Paulo, Johannesburg and Sydney. Continuing along these lines, Los Angeles, for example, would appear as

4. This produces a terrain within which these discontinuities can be reconstituted in terms of economic operations whose properties are not merely a function of the spaces on each side (i.e., a reduction to the condition of dividing line) but also, and most centrally, of the discontinuity itself, the argument being that discontinuities are an integral part, a component, of the economic system.

located on a variety of global circuits (including bi-national circuits with Mexico) which would be quite different from those of New York or Chicago. And a city like Caracas can be shown to be located on different circuits than those of Bogota.

This brings to the fore a second important issue. We can think of these cities or urban regions as criss-crossed by these circuits and as partial (only partial!) amalgamations of these various circuits. As I discuss later, some of the disadvantaged sectors in major cities today are also forming lateral cross-border connections with similarly placed groups in other cities. These are networks that while global do not run through a vertically organized framing as does, for instance, the network of affiliates of a multinational corporation or the country specific work of the IMF. For the city, these transnational circuits entail a type of fragmentation that may have always existed in major cities but has now been multiplied many times over. Topographic representations would fail to capture much of this spatialization of global economic circuits, except, perhaps, for certain aspects of the distribution/transport routes.

Sited Materialities and Global Span

It seems to me that the difficulty analysts and commentators have had in specifying or understanding the impact of digitization on cities results from two analytic flaws. One of these (especially evident in the U.S.) confines interpretation to a technological reading of the technical capabilities of digital technology. This is fine for engineers. But when one is trying to understand the impacts of a technology, such a reading becomes problematic. [5] A purely technological reading of technical capabilities of digital technology inevitably leads one to a place that is a non-place, where we can announce with certainty the neutralizing of many of the configurations marked by physicality and place-boundedness, including the urban. [6]

The second flaw, is a continuing reliance on analytical categorizations that were developed under other spatial and historical conditions, that is, conditions preceding the current digital era. Thus the tendency is to conceive of the digital as simply and exclusively digital and the non-digital (whether represented in terms of the physical/material or the actual, all problematic though common conceptions) as simply and exclusively that. These either/or categorizations filter out the possibility of mediating conditions, thereby precluding a more complex reading of the impact of digitization on material and place-bound conditions.

One alternative categorization captures imbrications (for a full development of this alternative see Sassen 2009: chapters seven and eight). Let me illustrate using the case of finance. Finance is certainly a highly digitized activity; yet it cannot simply be thought of as exclusively digital. To have electronic financial markets and digitized financial instruments requires enormous amounts of materiel, not to mention people. This materiel includes conventional infrastructure, buildings, airports, and so on. Much of this materiel is, however, inflected by the digital. Conversely, much of what takes place in cyberspace is deeply inflected by the cultures, the material practices, the

5. An additional critical issue is the construct technology. One radical critique can be found in Latour, and his dictum that technology is society 'Made Durable' (Latour 1991; 1996). My position on how to handle this construct in social science research is developed in Sassen (2009: ch 7). More generally see Mansell et al 2009.

6. Another consequence of this type of reading is the old practice of assuming that a new technology will ipso facto replace all older technologies that are less efficient, or slower, at executing the tasks the new technology is best at. We know that historically this is not the case. For a variety of critical examinations of the tendency towards technological determinism in much of the social sciences today see Wajcman 2002; Howard and Jones 2004; for particular applications that make legible the limits of these technologies in social domains see, e.g. Callon 1998; Avgerou, Ciborra and Land (2004); Cederman and Kraus 2005; for cities in particular see Graham 2004.

imaginaries, that take place outside cyberspace. Much, though not all, of what we think of when it comes to cyberspace would lack any meaning or referents if we were to exclude the world outside cyberspace. In brief, the digital and the non-digital are not exclusive conditions that stand outside the non-digital. Digital space is embedded in the larger societal, cultural, subjective, economic, imaginary structurations of lived experience and the systems within which we exist and operate. [7]

The complex imbrications between the digital (as well as the global) and the nondigital bring with them a destabilizing of older hierarchies of scale and often dramatic rescalings. As the national scale loses significance, along with the loss of key components of the nation-state's formal authority over the national scale, other scales gain strategic importance. Most especially among these are subnational scales, such as the global city, and supranational scales, such as global markets or regional trading zones. Older hierarchies of scale, emerging in the context of the ascendance of the nation-state, which continue to operate, are typically organized in terms of institutional size: from the international, down to the national, the regional, the urban, and the local. Today's rescaling cuts across institutional size and, through policies such as deregulation and privatization, cuts across the encasements of territory produced by the formation of nation-states. This does not mean that the old hierarchies disappear, but rather that rescalings emerge alongside the old ones, and that they can often trump the latter.

These transformations, which continue to entail complex imbrications of the digital and nondigital and between the global and the nonglobal, can be captured in a variety of instances. For example, much of what we might still experience as the "local" (an office building in our neighborhood) is actually something I would rather think of as a "microenvironment with global span," insofar as it is deeply internetworked. Such a microenvironment is in many senses a localized entity, something that can be experienced as local, immediate, proximate, and hence captured in topographic representations. It is a sited materiality.

But it is also part of global digital networks, which give it immediate far-flung span. To continue to think of this as simply local is not very useful or adequate. More important, the juxtaposition between the condition of being a sited materiality and having global span captures the imbrication of the digital and the nondigital and illustrates the inadequacy of a purely technological reading of the technical capacities of digitalization, which would lead us to posit the neutralization of the place-boundedness of that which precisely makes possible the condition of being an entity with global span. And it illustrates the inadequacy of a purely topographical reading.

A second example is the bundle of conditions and dynamics that marks the model of the global city. Just to single out one key dynamic: the more globalized and digitalized the operations of firms and markets, the more their central management and coordination functions — and the requisite material structures — become strategic. It is precisely because of digitalization that the simultaneous worldwide dispersal of operations, whether factories, offices, or service outlets, and system integration can

7. There is a third variable that needs to be taken account of when addressing the question of digital space and networks, though it is not particularly relevant to the question of the city. It is the transformations in digital networks linked both to certain technical issues and the use of these networks. (For critical accounts, see, e.g. Dean et al. 2007; Becker and Stalder 2009; Lovink 2007; Rogers 2004; MacKenzie and Wajcman 1999),

be achieved. And it is precisely this combination that raises the importance of central functions, which are largely placebound. Global cities are strategic sites for the combination of resources necessary for the production of these central functions. [8]

Much of what is liquefied and circulates in digital networks and is marked by hypermobility remains physical in some of its components. Take, for example, the case of real estate. Financial services firms have invented instruments that liquefy real estate, thereby facilitating investment and circulation of these instruments in global markets. Yet, part of what constitutes real estate remains very physical. At the same time, however, that which remains physical has been transformed by the fact that it is represented by highly liquid instruments that can circulate in global markets. It may look the same, it may involve the same bricks and mortar, it may be new or old, but it is a transformed entity. We have difficulty capturing this multi-valence through our conventional categories: if it is physical, it is physical; and if it is liquid, it is liquid. In fact, the partial representation of real estate through liquid financial instruments produces a complex imbrication of the material and the dematerialized moments of that which we continue to call real estate.

Cities as Frontier Zones: The Formation of New Political Actors

A very different type of case can be found in the growth of electronic activism by often poor and rather immobile actors and organizations. Topographic representations that describe fragmentations, particularly the isolation of poor areas, may well obscure the existence of underlying interconnections. What presents itself as segregated or excluded from the mainstream core of a city can actually be part of increasingly complex interactions with other similarly segregated sectors in cities of other countries. There is here an interesting dynamic where top sectors (the new transnational professional class) and bottom sectors (e.g. immigrant communities or activists in environmental or anti-globalization struggles) partly inhabit a cross-border space that connects particular cities.

Major cities, especially if global, contain multiple low-income communities many of which develop or access various global networks. Through the Internet, local initiatives become part of a global network of activism without losing the focus on specific local struggles. It enables a new type of cross-border political activism, one centered in multiple localities yet intensely connected digitally. This is in my view one of the key forms of critical politics that the Internet can make possible: A politics of the local with a big difference — these are localities that are connected with each other across a region, a country or the world. [9] Because the network is global does not mean that it all has to happen at the global level.

But also inside such cities we see the emergence of specific political and subjective dimensions that are difficult to capture through topographic representations. Neither the emergence nor the difficulty are new. But I would argue that there are times where both become sharper — times when traditional arrangements become unsettled. Today is such a time. Global cities

8. There are other dimensions that specify the global city; see Sassen, The Global City.

9. I conceptualize these "alternative" circuits as countergeographies of globalization because they are deeply imbricated with some of the major dynamics constitutive of the global economy yet are not part of the formal apparatus or of the objectives of this apparatus. The formation of global markets, the intensifying of transnational and trans-local business networks, the development of communication technologies which easily escape conventional surveillance practices --all of these produce infrastructures and architectures that can be used for other purposes, whether money laundering or alternative politics.

become a sort of new frontier zone where an enormous mix of people converge and new forms of politics are possible. Those who lack power, those who are disadvantaged, outsiders, discriminated minorities, can gain presence in global cities, presence vis-à-vis power and presence vis-à-vis each other. This signals, for me, the possibility of a new type of politics centered in new types of political actors. It is not simply a matter of having or not having power. There are new hybrid bases from which to act.

Digital networks are contributing to the production of new kinds of interconnections underlying what appear as fragmented topographies, whether at the global or at the local level. Political activists can use digital networks for global or non-local transactions and they can use them for strengthening local communications and transactions inside a city or rural community. Recovering how these new digital technologies can serve to support local initiatives and alliances across a city's neighborhoods is extremely important in an age where the notion of the local is often seen as losing ground to global dynamics and actors and digital networks are typically thought of as global. What may appear as separate segregated sectors of a city may well have increasingly strong interconnections through particular networks of individuals and organizations with shared interests. Any large city is today traversed by these "invisible" circuits.

REFERENCES

— Avgerou, Chrisanthi, Claudio Ciborra and Frank Land. 2004. The Social Study of Information and Communication Technology Innovation, Actors, and Contexts. Oxford, UK: Oxford University Press.

— Becker, Konrad and Felix Stalder (eds). 2009. Deep Search. The Politics of Search beyond Google. New Jersey: Transaction Publishers.

— Callon, Michel. 1998. The Laws of the Markets. Oxford, UK: Blackwell Publishers.

— Cederman, Lars-Erik and Peter A. Kraus. 2005. "Transnational Communications and the European Demos." Pp. 283-311 in Digital Formations: IT and New Architectures in the Global Realm, edited by Robert Latham and Saskia Sassen. Princeton, NJ: Princeton University Press.

— Dean, J, Anderson, J. W., and Lovink, G. (2006) Reformatting Politics: Information Technology and Global Civil Society. London: Routledge.

— Graham, S. (ed). 2004. Cybercities Reader. London: Routledge.

— Howard, P.N. and Jones, S. (eds) (2004) Society Online: The Internet in Context, London:Sage.

— Latour, Bruno. 1991. "Technology Is Society Made Durable." In A Sociology of Monsters, edited by John Laws. London: Routledge.

— ——, 1996. Aramis or the Love of Technology. Cambridge, MA: Harvard University Press.

— Lovink, Geert. 2007. Zero Comments: Blogging and Critical Internet Culture.

— MacKenzie, Donald and Judy Wajcman. 1999. The Social Shaping of Technology. Milton Keynes, UK: Open University Press.

— Mansell, Robin, Chrisanthi Avgerou, Danny Quah, and Roger Silverstone (eds) 2009, The Oxford Handbook of Information and Communication Technologies (Oxford: Oxford University Press).

— Rogers, Richard. 2004. Information Politics on the Web. Cambridge, MA: MIT Press.

— Sassen, Saskia. 2001. The Global City: New York, London, Tokyo. Princeton, NJ: Princeton University Press.

— ——, 2009. Territory, Authority, Rights: From Medieval to Global Assemblages. NJ: Princeton University Press.

— ——, 2010. "When the city itself becomes a technology of war." Theory, Culture, Society. Forthcoming.

— Wajcman, Judy (ed). 2002. "Special Issue: Information Technologies and the Social Sciences." Current Sociology 50(3).

THE URBAN CULTURE OF SENTIENT CITIES: FROM AN INTERNET OF THINGS TO A PUBLIC SPHERE OF THINGS

MARTIJN de WAAL

At certain points in the history of architecture and urban planning, the disciplinary debate on how to apply new technologies surpasses the boundaries of the professions involved. At those times, the hopes and fears found in the disputes between architects, policy makers, engineers and planners are extended to a broader discussion about urban and societal change. Then, the central issue is not merely how to solve a specific spatial problem or improve a construction method with the help of a new technology. Rather, the debate revolves around its possible impact on urban society at large. What does this new technology mean for urban culture, what impact does it have on how we shape our identities and live together in the city? When those questions surface, Dutch philosopher René Boomkens argues, the professional debate has turned 'philosophical'. [1]

The discourse on 'Sentient Cities' that has arisen over the last few years can be understood as such a philosophical enterprise. [2] What is at stake in the debate is not so much the issue of how to engineer smarter buildings that sense — and adapt to — our daily routines or idiosyncratic preferences. Rather, our in-car navigators, friend finding 'solutions', location based information systems and other urban sensing technologies may very well force us to rethink some of the core concepts through which we understand and value urban life.

Here I will show that the debate about the Sentient City can be understood as a dispute concerning the urban public sphere. On the one hand, the rise of sentient technologies is said to contribute to the (already on-going) demise of urban public spaces such as town squares, multifunctional streets and public parks. On the other hand, there is a hope that those same sentient technologies could enable new forms of publicness and exchange. These are no longer based on bringing people with different backgrounds and opinions spatially together (as in coffeehouses or town squares), but on the organization of publics around particular issues of concern.

The Sentient City

Before I delve into the debate on the relation between sentient technologies and the urban public sphere, I first want to spend a few lines on the term 'Sentient City' itself. What exactly do we mean when we invoke the emergence of a 'Sentient City'? The artist and architect Mark Shepard puts it this way. He states that increasingly, it is the 'dataclouds of 21st century urban space' [3] that shape our experience of the city. All over the city, 'intelligent' applications have started sensing what is happening around them and reacting to it — be it smart traffic lights or cctv camera's whose images are computer analyzed for suspicious behavior. Add to this the increase of tracking devices such as cell phones that most urbanites carry, and as a result the city has become 'sentient'.

Now of course it is not the city itself that perceives or even is sentient, but rather the combined apparatus of tracking and sensing devices — operated by different actors — that note what is going on in the city and output their impressions in all sorts of data streams. Neither is this emergence of the sentient city a singular movement driven by a centralized bureaucracy

1. R. Boomkens, Een Drempelwereld: Moderne Ervaring En Stedelijke Openbaarheid (Rotterdam: NAi Uitgevers, 1998).

2. Over the last few years, the term 'sentient city' has come up in a number of publications, exhibitions and events. For instance Stephen Graham and Mike Crang wrote an article in 2007 in which they speak of 'environments that learn and possess anticipation and memory' and relate this vision to three different takes on the city, varying from 'market-led visions of customized consumer worlds', 'military plans for profiling and targeting' and 'artistic endeavours to re-enchant and contest the urban informational landscape of urban sentience'. The term is also used in the title of the exhibition Toward the Sentient City, organized by the Architectural League in New York in the fall of 2009 and curated by Mark Shepard. Here the theme and framing of the exhibition are related to a series of publications called the 'Situated Technologies Pamhplets', published by the Architectural League of New York. In another example the term 'Sentient City' is sometimes also used by the SENSEable City Lab from MIT, a research institution that 'explores the "real-time city" by studying how distributed technologies can be used to improve our understanding of cities and create a more sustainable ways of interacting in urban environments.' The term is related to similar labels that also describe the increasing role of computing technologies in the constitution of everyday urban life, such as The Real Time City, Urban Informatics, Urban Computing and others. Each stems from its own

or company, established at a single address to which one could send a letter of complaint or e-mail a feature request. The Sentient City should be understood as a collection of plural research traditions, performed and commissioned by divergent actors all with their own motivation and implicit understanding of what a city is or should be. They vary from government agencies that want to bring order to city space, politicians that would like to promote citizenship, companies that want to offer personalized services, community workers that hope to promote solidarity or mutual understanding, artists that want to criticize consumer culture and urbanites who may embrace, adapt or reject some or other of these offerings.

The concept of the Sentient City is not an arbitrarily chosen stock term for these developments. In a definition drawn up by Mark Shepard, he explicitly refers to the Latin roots of this term to explain what he means with that term: 'Sentience refers to the ability to feel or perceive subjectively, and does not necessarily include the faculty of self-awareness.' [4] This emphasis on subjectivity foregrounds the fact that the data streams generated by the Sentient City may seem like instances of objective fact gathering, whereas in reality they are far from it. For starters, the decision regarding which data to collect and which to ignore and how to classify it, is already a highly political choice. Next, the data generated by the Sentient City is interpreted by software algorithms and actuation devices, and there is nothing objective about that either: it is a highly normative process, where subjective values, legal codes and power relations are turned into software code on the base of which sentient technology decides, acts and discriminates. [5]

This foregrounding of the normative side of the Sentient City goes against the grain of the discourses of 'ubiquitous computing' (or ubicomp) and 'urban computing' that play a dominant role in the debate on the Sentient City. In ubicomp, an application is usually thought successful if it makes the computer disappear. While we carry on our daily routines, computation technology — calmly operating in the background — will make our live more easy, efficient or exciting — whatever way we would want it. Not only does it do away with the need to interact with those beige boxes on our desktops (which of course is not a bad thing per se), it also renders invisible and presents as natural the visions of what a city should be and for whom these social interventions are enabled. The conceptualization of the Sentient City can thus be understood as a deliberate move in the debate on the role of computers in urban society.

The Sentient City, Urban Culture and the Public Sphere

In many of the debates that foreground the possible impact of sentient technology on urban culture, sentient technologies are linked up to the role of the public sphere in urban society. The quintessential characteristic of urban life, as urban theory since Simmel has pointed out, is that urbanites are to live together with strangers who not only will remain strangers, but may also have a completely different outlook on life. Yet somehow, all citizens have to find a way to work things out. The public sphere plays an important role in this. It is here that strangers are

disciplinary modus operandi and brings a different approach to computing and urban society to the table.

3. Mark Shepard, "Curatorial Statement," The Architectural League, http://www.sentientcity.net/exhibit/?p=3.

4. Ibid.

5. See for instance Stephen Graham, "Software-Sorted Geographies," Progress in Human Geography 29, no. 5 (2005)., Steven Graham and Mike Crang, "Sentient Cities : Ambient Intelligence and the Politics of Urban Space.," Information, communication & society 10, no. 6 (2007)., Nigel Thrift and Shaun French, "The Automatic Production of Space," Transactions of the Institute of British Geographers 27, no. 3 (2002).

confronted with each other, become aware of one another and have to come to terms with each other. [6]

The most famous proponents of this notion of the public sphere are Hannah Arendt, Richard Sennett and Jürgen Habermas. Although their exact positions differ, the central idea is that a society needs a place in which these differences are brought together. The notion of a public sphere is contrasted to the private sphere. In an ideal public sphere, participants are able to distance themselves from their private identities and focus on a common interest. [7] As Frei and Böhlen have pointed out, the public sphere in an Arendtian sense is 'the site of collective performance that brings together those who are different from one another precisely because they are different. The collective that acts in the public realm is not a uniform entity such as a class, a nation, or a mass. What brings people together here is exactly what separates them from each other; in other words, according to Arendt, the public realm is like parentheses that hold together the differences between people.' [8] Most theories describe the urban public sphere as a physical site, it consists of actual, physical places where people are confronted with one another. Although Habermas interestingly notes that already in the 17th Century media did play an important role as well. In the Coffee Houses that for him were the quintessential example of the emerging public sphere of that time, newspapers played an important role. They were sometimes read aloud and discussants would often send letters to the editor after they had discussed the articles over a cup of coffee, or *in real life* as we would say today. This way the discussion could be continued in other coffee houses, with the newspaper forging the link between the instances in which a public sphere came into being during the conversations in the coffee houses. [9]

This ideal of the public sphere has been said to lie under attack ever since it has been conceived. Privatization, parochialization and intimization are the main culprits, or more concretely: the suburb, the automobile, the television and — in more recent years — the mobile phone. Both Sennett and Arendt have argued that the new wealth of the middle class has enabled them to segregate themselves socially, to physically surround themselves with people of their own liking, and thus retract from public life into their own parochial domains. This may in due time erode the capacity to empathize with others and the solidarity necessary to upkeep an inclusive urban society. [10]

The Sentient City as Threat or Savior of the Public Sphere

Some theorists fear that the affordances of sentient technologies reinforce this demise of this public sphere. Many sentient applications that are currently in development are based on the implicit idea of the city as a collection of services and infrastructures to be managed as efficiently as possible. Alternatively they offer personalized versions of the city through search and 'discovery devices'. These latter services follow users' whereabouts through the city and use that information to draw up a profile of every user. These profiles are compared to each other and used to make recommendations to visit a restaurant or a bar. The goal

6. Amanda Williams, Erica Robles, and Paul Dourish, "Urbane-Ing the City: Examining and Refining the Assumptions Behind Urban Informatics," in Handbook of Research on Urban Informatics: The Practice and Promise of the Real-Time City, ed. Marcus Foth (Hershey, New York, London: Information Science Reference, 2008)., Boomkens, Een Drempelwereld : Moderne Ervaring En Stedelijke Openbaarheid.

7. Boomkens, Een Drempelwereld : Moderne Ervaring En Stedelijke Openbaarheid.,

8. Hans Frei and Marc Böhlen, Situated Technologies Pamphlet 6: Micropublicplaces, ed. Omar Khan, Trebor Scholz, and Mark Shepard, Situated Technologies Pamphlets (New York: The Architectural League of New York, 2010)., Jurgen Habermas, The Structural Transformation of the Public Sphere. An Inquiry into a Category of Bourgeois Society (Cambridge, MA: MIT Press, 1991)., Hannah Arendt, The Human Condition (Chicago: University Of Chicago Press, 1958)., Richard Sennett, The Uses of Disorder: Personal Identity and City Life (New York: Norton, 1970); The Fall of Public Man (New York: Knopff, 1977).

9. Habermas, The Structural Transformation of the Public Sphere. An Inquiry into a Category of Bourgeois Society.

10. Sennett, The Uses of Disorder : Personal Identity and City Life , Frei and Böhlen, Situated Technologies Pamphlet 6: Micropublicplaces.

is to have the user 'discover' places in the city that are both new to him, and where he can immediately feel at home at the same time. Other initiatives depart from control and security-issues: they use sentient technology to prevent potential unrest or to allow or deny access to certain users.

Sensing technologies thus have the affordance to sense who or what is near them and filter this data according to the preferences of its users. For them the city may turn into a patch network of parochial spaces. If they live up to their promises, these technologies promise that urbanites never have to leave the comfort of being surrounded by like minded people. The other way around: access to certain urban places might only be given to authorized people recognized by embedded sensors.

Combined, in a dystopian scenario, these appropriations of the technology might contribute to what Belgian philosopher Lieven de Cauter has called a 'capsular society' — a city of privatized capsules with different functions — dwelling, shopping, consuming accessible only to those with the right RFID-chip in their wallet. [11]

There are also more optimistic accounts. As Stephen Graham and Mike Crang amongst others have pointed out, many artists have embraced locative media to re-activate the urban public sphere. For instance geoannotation (software through which people can mark-up particular urban places with stories, photo's or video) makes it possible for passers-by to 'sense' alternative stories, points of view or issues related to the places they visit. The idea behind many of these projects is to have urban space function once more as a site of exchange and communication between citizens. Only this time, they don't have to be there physically present at the same time. They can learn about the visions and stories of other citizens who have passed by earlier through locative media annotation services. [12]

Toward a different public sphere?

What both critical and optimistic scenarios mentioned above have in common is that they still depart from the idea of the urban public sphere as a physical site in which differences are to be brought together. There are also a number of theories and art projects that point to an alternative and new conceptualization of the public sphere. In this vision, the public does not come together in a physical site, but rallies around an issue of concern, that is raised through sentient technologies.

One of the clearest descriptions of this shift from *public spheres* to *publics* can be found in Frei and Böhlen's previously mentioned pamphlet. They draw on Latour's Parliament of Things, a theory that in turn builds upon Dewey's notion of issue publics: publics that (temporarily) form around specific issues in which they have taken a certain ownership. This public assembles — as in Arendt's notion of the public sphere— not because everyone agrees but because they disagree and need to come to terms in some way or another. [13]

Publics can form around shared issues of concern for which people feel some form of ownership. The infrastructure of the sentient city itself can form such a shared issue of concern. Frei and Böhlen describe a number of ways in which 'micropublics'

11. Lieven De Cauter, De Capsulaire Beschaving. Over De Stad in Het Tijdperk Van De Angst (Rotterdam: NAi Publishers, 2004). see also Marc Schuilenburg and Alex De Jong, Mediapolis (Rotterdam: 010 Publishers, 2006).

12. Graham and Crang, "Sentient Cities: Ambient Intelligence and the Politics of Urban Space..", see also Lily Shirvanee, "Locative Viscosity: Traces of Social Histories in Public Space," Leonardo Electronic Almanac 13, no. 3 (2006), http://leoalmanac.org/journal/vol_14/lea_v14_n03-04/toc.asp., Anne Galloway, "A Brief History of the Future of Urban Computing" (Carleton University, 2008)., Williams, Robles, and Dourish, "Urbane-Ing the City: Examining and Refining the Assumptions Behind Urban Informatics."

13. Frei and Böhlen, Situated Technologies Pamphlet 6: Micropublicplaces., Noortje Marres, "Zonder Kwesties Geen Publiek," Krisis, no. 2 (2006)., Bruno Latour, "From Realpolitik to Dingpolitik: An Introduction," in Making Things Public: Atmospheres of Democracy, ed. Bruno Latour and P Weibel (Cambridge MA: MIT Press, 2005).

might form around communal urban infrastructures or institutions such as schools, parks, water plants. Whereas these are all conceptual blueprints, Laura Forlano points out that such publics did emerge around a number of wireless networks she has studied in several cities around the world. They succeeded — at least temporarily — in bringing people with different identities and backgrounds together. [14]

Publics may also gather in a different way. One of the promises of the rise of sentient technologies is that things, objects and issues can record their own 'biographies.' [15] The project Trash Track by MIT's SENSEable City Lab demonstrates this affordance. For this exhibit, trash items such as paper cups are tagged with a GPS-device and mobile phone modem. After it has been disposed of, the item sends text messages with its location, so we can follow its track from recipient to waste disposal site. The hope expressed through this project is that knowing will lead to a change in doing: the fact that we know where our trash ends up should make us more aware of the problem we create by throwing things away.

In a published conversation Bratton and Jeremijenko point out that there is a lot of hope that the data gathered by the sentient city will lead to engagement with important issues of our times. The collection and visualization of data about environmental pollution might become a 'thing' — an issue of concern — around which a public might assemble. [16] Similarly, Laura Forlano points out that with new sensing technologies, it becomes possible to point your mobile phone (for instance) at a product in the supermarket and immediately learn whether it has been manufactured in a sweat-shop or in an environmentally unfriendly way. [17]

It is, however, not as simple as that. A beautiful visualization of data gathered by sentient technologies might be just that — an aesthetically pleasing work of art decorating a museum wall. Bratton and Jeremijenko point out that many of such mappings do not really lead to the formation of active publics. They don't change who is asking what kind of questions, they do not show alternatives, or give anyone a sense of agency.

"And so, do these projects change who is asking the questions? Are these designers now asking the question of how this pollutant is made, who made it, where is it coming from, where is it going, what do we do about it, or not? ... Who collected [the data] and under what conditions. That is, what does the data actually represent?" [18]

It makes a difference whether the information received about issues is collected by marketing agencies or institutions that are by law obliged to register and publish such data, or whether a group of local activists who have a completely different interest in the issue of concern has programmed the sensors and algorithms involved. In short, Bratton and Jeremijenko argue, we do not need mere mappings, we need *interfaces* that allow the public not just to take note of a dataset, but that also provide it with an agency to actually get involved.

As Bratton has observed, our urban societies are no longer restricted to their administrative territory, bounded by the dashed lines on the map. Instead, we live in cities 'where flows

14. Laura Forlano and Dharma Dailey, "Community Wireless Networks as Situated Advocacy," in Situated Technologies Pamphlets 3: Situated Advocacy, ed. Omar Khan, Trebor Scholz, and Mark Shepard, Situated Technologies Pamphlets (New York: The Architectural League of New York, 2008).
15. Bruce Sterling, Shaping Things (Cambridge, MA: MIT Press, 2005).
16. Benjamin Bratton and Natalie Jeremijenko, "Suspicious Images, Latent Interfaces," in Situated Technologies Pamphlets 3: Situated Advocacy, ed. Omar Khan, Trebor Scholz, and Mark Shepard, Situated Technologies Pamphlets (New York: The Architectural League of New York, 2008).
17. Forlano and Dailey, "Community Wireless Networks as Situated Advocacy."
18. Bratton and Jeremijenko, "Suspicious Images, Latent Interfaces."

move in and out of geographies, where territories are occupied by multiple collectives at once, and where the procedures, networks, and assemblages of objects and things are vastly distanced from our own capacities to perceive them.' [19] It is exactly for that reason that bringing people together spatially may no longer be a viable idea for maintaining a public sphere. Rather, we should start thinking about how we can move from an 'Internet of things' to a public sphere centered around things.

That is not just a philosophical shift. It is also a practical matter. How to design interfaces that go beyond mere mapping of things could become one of the most important design challenges of our times for everyone concerned with the role of the public sphere in a democratic society.

19. Ibid.

SPACE, FINANCE, AND NEW TECHNOLOGIES

KAZYS VARNELIS

The framers of the Sentient City exhibit suggest that in the contemporary urban environment, informatic systems play as big — if not greater a role — as the familiar forms of the buildings that surround us. Indeed, no matter how compelling the series of alternative future trajectories of technology that the show features, aren't our cities already sentient? Pervasive wireless Internet, smart phones, locative media, augmented reality, and embedded sensors aren't science future anymore, rather they're part of everyday life, exciting only for the few minutes after they are introduced.

If they are banal now, these technologies played a crucial role in the development of a new spatial regime: mobile phones, initially only in the hands of a wealthy élite, spread to the majority of the world's population while, in the developed world, the Internet became both the dominant form by which most individuals received media, as well as a prime means of communication, work, and social networking. Processes of globalization played a key role too, making collaborative businesses, work across distance and frequent travel common in the service sector. Our sense of place has not so much become undone as transformed, dispersing into a multidimensional spatiality that, for the first time, is not an abstract matter but rather part of lived experience.

Going against network culture's ahistorical grain to situate our contemporary spatiality within a broader historical frame allows us to better understand the change.[1] As a historical framework for discussing space, Henri Lefebvre's The Production of Space still seems to me to be the most workable. In it, Lefebvre identifies three successive spatial regimes: absolute space, historical space, and abstract space. [2] In the regime of absolute space, humans value spaces for their natural qualities, defining them as sacred (the Acropolis, the sacred spring, the sacred forest), only to obliterate their natural characteristics with constructions and interventions. Historical space evolves out of absolute space, as humans value spaces that have been the object of habitation and events (e.g. the town, the shrine at a saint's birthplace). The last of the three, abstract space, emerges when humans quantify and mathematize territory, assigning value through capitalist and bureaucratic organizations. Throughout, spatial regimes build upon spatial regimes. Thus the spring from which a local people drew fresh water becomes sacred and a temple is built upon it. When the people are converted, Christians obliterate the pagan temple, build a cathedral on the site, and centuries later wonder why the foundations are sinking. By then, however, the buildable area facing the square in front of the Cathedral is the object of increasing speculation by land developers.

In setting out his approach, Lefebvre argues that "(social) space is a (social) product," that space is not a cultural superstructure determined by a mode of production, but rather is a construction that is both produced within a society and serves to reproduce that society. Just as a society's conception of space is influenced by the way that economic forces shape it, it is also an agent of its own that impacts the development of such forces. [3] So, if in modernity, the Cartesian grid gives value to land,

1. For atemporality and ahistoricity under network culture as well as a framing argument for the position this essay, see the book that I am currently drafting, Life After Networks: A Critical History of Network Culture online at http://varnelis.net/network_culture
2. Henri Lefebvre, The Production of Space, trans. Donald Nicholson-Smith, La production de l'espace copyright Editions Anthropos 1974, 1984 ed. (Oxford, UK and Cambridge, Massachusetts: Basil Blackwell Ltd., 1991), 47-53.
3. Lefebvre, 26.

the claim to map the world accurately emerges first not in the economy but rather in art and science, through the invention of perspective, the Cartesian grid, and grid-based cartography.[4]

Abstract space, Lefebvre writes, seeks to subordinate all spatial models to its inexorable, mathematical logic.[5] But abstract space is a process, not an end point; Lefebvre observes that rather than being a homogeneous condition, abstract space is the process of creating spatial homogenization, producing a form of space based on value.[6] Charlie Gere elaborates on abstraction as a process, pointing out that in making the world exchangeable, it is fundamental for investment, trade, and management. Abstraction makes possible the interchangeability and interoperability of machines and humans. In doing so, it unmoors objects from their contexts, allowing them to circulate freely, traded for their exchange-value.

Abstract space is embodied in the metropolis, the primary site of production for capitalism. There, a people come together into close quarters for ever-greater efficiency. In the metropolis, sociologist Ernest Burgess observes, newcomers undergo a process of disorganization in which their habitual ways are undone prior to their experiencing a reorganization that makes them productive. This process allows immigrants from foreign lands and the countryside to become metropolitan subjects, but it also produces value for the city through the introduction of new ideas and new energies.[7] But Georg Simmel, in "The Metropolis and Mental Life," also observes that faced with "the intensification of nervous stimulation," the metropolitan individual shuts down, becoming blasé or indifferent to the surroundings. Simmel's diagnosis parallels that of contemporary psychologists, such as George Beard, who in 1869 identifies the stresses of urban life as causing kinesthetic neuresthenia, an overstimulation forcing individuals to shut down, becoming apathetic, depressed, and withdrawn.[8]

The metropolitan subject is fundamentally transformed from its rural predecessor. Instead of life-long, emotionally deep relationships with a few individuals, the modern subject maintains relationships with a large number of individuals, treating relationships superficially and managing them with the intellect. Anonymous exchanges governed by money ensure that they will stay matter-of-fact, subject to reason not emotion. Strict punctuality, meticulous calculation and preciseness in all things assure the function of this system.[9]

For Lefebvre, the stultifying effects of this over-rationalized life produce a contradiction: where the public life of the bourgeois individual is one of blasé anonymity, formality, rationality, and money relationships, private life centers around the family and the domestic interior, spaces of reproduction intended as refuges to the homogenizing, reifying space of the city.[10] These interiors, Susan Sidlauksas writes, are a "metaphor for bourgeois identity," representing the uniqueness of the self against the crushing nature of the metropolis.[11]

But read in Foucauldian terms, rather than a contradiction, the interior is part of the era's disciplinary logic. The process of creating subjects capable of regulating themselves in a modern society requires appropriate regulation, which takes place in

4. To be clear, the grid also emerged because of needs developed in earlier forms of land valuation, for example Alberti's map of Rome was for the use of Pope Nicholas V—or in military technologies. Compare with Lefebvre's discussion of the invention of perspective, which he attributes to a space emerging out of the Tuscan reconstruction of the relationship of the town and the country. Lefebvre, 78.

5. Lefebvre, 49-50.

6. Lefebvre, 287.

7. Robert Ezra Park et al., The City, University of Chicago Sociological Series. (Chicago: The University of Chicago press, 1928), 54.

8. Anson Rabinbach, The Human Motor: Energy, Fatigue, and the Origins of Modernity (New York: Basic Books, 1990), 153.

9. Georg Simmel, David Frisby, and Mike Featherstone, Simmel on Culture: Selected Writings, Theory, Culture & Society (London: Sage Publications, 1997), 174-86.

10. Lefebvre, The Production of Space, 51-52.

11. Susan Sidlauskas, "Psyche and Sympathy: Staging Interiority in the Early Modern Home'," in Not at Home: The Suppression of Domesticity in Modern Art and Architecture, ed. Christopher Reed (London: Thames and Hudson, 1996), 65-80.,

a series of enclosures: factories, schools, barracks, hospitals, offices, as well as the bourgeois interior, all subjected to the dominating force of inspection and the gaze which enforces its power over space.[12] If in the first group the gaze is external, in the home regulation happens from within, with the patriarch as the crux between the home and the world, the regulating force ensuring order.

Technology was crucial to the new sense of space. For one, telecommunications made it possible for the metropolis to be the site of industrial production and management, enabling the coordination of shipments, remote management of resources (to a degree), and transmission of information worldwide. The growth of the city core or downtown was made possible by — and produced in part by — those new technologies which gave rise to a managerial core separate from outlying areas of production. But everyday life was also colored by a growing sense of simultaneity that space-time compression produced. With the invention of the telegraph and wireless radio, the moderns experienced the world as one for the first time. Within a day, events such as the Japanese defeat of the Russian Fleet at Tsushima in 1905 and the loss of the Titanic in 1912 were topics of discussion worldwide. But no matter how unnerving or liberating this sense of simultaneity, for most people it was largely a third-hand matter, experienced through the intervening media of newspapers, magazines, radio, or eventually television. Even if radio and television made it possible for individuals to listen to events taking place in real time, throughout modernity everyday life was still confined to one's surroundings. Long-distance telephony, for example, was rarely used for personal reasons well into the 1970s and international telephony only took off in the 1990s.[13]

The twentieth century would bring new pressures to the home: the radio and television replaced the piano and the fireplace as gathering points for the family. Individuals were subsumed into the mass, addressed by a media that perceived its audience as homogeneous, consumers rather than producers. These top-down, one-to-many media created temporary reconciliation but also left many with a feeling of emptiness afterward. Connection would be simulated, but remained one-way.

During the Great Depression, over-investment in real estate posed a problem to the sustainability of cities. Building stock in urban cores was expensive even as it aged and became obsolete. Development in such areas faced high land values and the need to still pay off existing, if out-of-date, structures, while the inefficiencies caused by congestion posed further difficulties.[14] In the United States, capital began to seek more lucrative prospects for development in outlying areas, both secondary business districts and suburbs. A constellation of forces — economic, strategic, managerial, and cultural — combined with increased automobility to spur the growth of both residences and workplaces in suburban areas starting in the postwar era. City cores began a decades-long process of decay from which all but a few have not recovered.

By the early 1960s, urban sprawl had grown to the point that geographer Jean Gottmann defined the continuous urban region from northern Virginia to southern New Hampshire as

12. Michel Foucault, Discipline and Punish:The Birth of the Prison, 1st American ed. (New York: Pantheon Books, 1977), 104-34.

13. See David Harvey, The Condition of Postmodernity: An Enquiry into the Origins of Cultural Change (Oxford, UK: Blackwell, 1989) and Stephen Kern, The Culture of Time and Space 1880-1918 (Cambridge, MA: Harvard University Press, 1983).

14. Robert M. Fogelson, Downtown: Its Rise and Fall, 1880-1950 (New Haven: Yale University Press, 2001), 183-218.

a megalopolis.[15] Other regions soon followed: in the United States, Southern California; in Europe, Amsterdam-Brussels-Hannover and London-Leeds-Manchester; in Japan the Tokaido Corridor. Especially in the United States, but also in other such regions worldwide, manufacturing shifted toward outlying areas, still part of metropolitan or megalopolitan regions but now part of postsuburbia. In this new territory individuals who resided in suburbs would commute, not to city cores, but to other suburbs.[16] As economies worldwide underwent a restructuring to post-Fordism, rigid and hierarchical forms of organization were replaced with more flexible structures. With less expensive land, an educated population less likely—particularly in the American South and Southwest — to be committed to unions, and a supply of abundant part-time labor in housewives seeking flexible employment, suburbs were apt territories for the expansion of a new generation of businesses. Decentralization of manufacturing soon spread further as industries turned to contractors and suppliers in developing countries to fulfill their orders. In addition, new models of corporate communications suggested that horizontal office buildings found outside city cores would be more communicationally efficient than vertical skyscrapers.[17]

But as Saskia Sassen explains, the resurgence of New York, and to a lesser degree, other regional centers in the 1980s was not at odds with this decentralization. Instead it was part of the same logic. As industrial production shifted further and further out of cities into larger metropolitan regions and to developing regions, global cities served as places for financial production and control. The cores of these cities would be marked by an often-deliberate evisceration of their industrial capacities. But the cores of global cities, possessing advanced telecommunication facilities and existing financial institutions wound up the new centers of global economic control.[18]

If we think of the global economic system — not just now but historically — as a network, then we shouldn't be surprised by the rise of the global cities model. Power-law effects tend to emerge in networks, with connected nodes becoming ever more highly connected. The global city is thus a kind of network effect. So we have to be cautious about the urban renaissance of the last two decades; it has worked, to a degree, in global cities like New York and London, but many other cities have seen long-term sustained population declines.[19]

Beginning in the 1960s, over accumulation of capital and declining profit rates have posed a problem for the world capitalist system, particularly in advanced economies. On the one hand, investors demanded ever greater returns, on the other hand, rates of return in manufacturing fell steadily. The solution was a growing turn to financialization, leading to the creation of ever-more-complex financial instruments. In traditionally more economically advanced countries — the United States and the United Kingdom, but also much of the EU and Japan — the result has been that finance has come to dominate the economy. Kevin Phillips, for example, sums it up, noting the "extraordinary rise of the U.S. financial sector from 11–12 percent of the gross national product back in the 1980s to a stunning 20–21 percent

15. Jean Gottmann, Megalopolis; the Urbanized Northeastern Seaboard of the United States (New York: Twentieth Century Fund, 1961).
16. Rob Kling, Spencer C. Olin, and Mark Poster, Postsuburban California:The Transformation of Orange County since World War II (Berkeley: University of California Press, 1991).
17. John F. Pile, Open Office Planning: A Handbook for Interior Designers and Architects (New York: Whitney Library of Design, 1978).
18. Saskia Sassen, The Global City: New York, London, Tokyo (Princeton, NJ: Princeton University Press, 1991).
19. Philipp Oswalt and Kulturstiftung des Bundes., Shrinking Cities, (Ostfildern: Hatje Cantz, 2005).

of the U.S. gross domestic product by 2004–2005. During that same quarter century, manufacturing, for a century the pillar of our economy, slipped from about 25 percent to just 12 percent." Nor has the recent economic crisis reversed the trend.[20]

Where abstract space still adhered to a Euclidean model of the world, with space being readily mappable in terms of a coordinate system, our own spatial regime exceeds that model. As Castells observes, the network organizes space according to the effects of the power law. He identifies new spatial logic as the "space of flows," that as "the material organization of time-sharing social practices that work through flows." The space of flows is comprised of electronic networks that allow the network economy to function together with the specific nodes and communication hubs on the network and the spatial manifestation of this logic for the business élite. The latter, he explains, takes place through social micro-networks that emerge in the sorts of places that the élite inhabit: exclusive restaurants, cultural events, clubs, and so on.[21] The space of flows, in other words, is already multidimensional, operating at simultaneous levels that cannot coherently map upon each other.

During the last two decades, financialization has added a new layer to our spatial regime. Although real estate speculation has been part of the metropolis since the start, more recently the securitization of real estate through instruments like Real Estate Investment Trusts and Mortgage-Backed Securities allowed investors to purchase fragmentary shares in real estate in exchange for income from mortgages or rents. The division of real estate into fragmentary shares, together with the creation of secondary or derivative instruments such as Credit Default Swaps purportedly allowed risk to be reduced through diversification. The confidence that this created made possible the management of more risky investments such as subprime loans to the poor and interest-only mortgages for under-capitalized house flippers. Space was turned into abstract financial instruments that could then circulate freely on the network, investment becoming a matter no longer just of realizing direct returns on the property, but rather on financial processes taking place in the market itself.

The result was the rapid rise of real estate worldwide to the point that values lost any economic sense. Where the worth of a given piece of real estate was traditionally tied to its worth derived from the price it could rent for, prices began to be valued for the logic of the market alone. Thus, market models widely used by investors and homeowners began to determine the price of real estate, allowing it to go wildly out sync with the traditional forms of valuation. With the market awash in liquidity and with a lifestyle of frequent — if not constant — travel common among the rich, it became increasingly common to purchase multiple pied-à-terres in global cities, helping to precipitate a rapid increase in price. Thus, in 2010, even after the crash, an apartment in New York city trades vastly in excess of its rent potential. Throughout the "new economy" that began in the 1990s and was revived under the real estate boom, there was an assumption that technological innovation had made productivity grow while an advancement

20. Kevin Phillips, Bad Money: Reckless Finance, Failed Politics, and the Global Crisis of American Capitalism (New York: Viking, 2009), xiii..
21. Manuel Castells, The Rise of the Network Society, 2nd ed. (Oxford, UK: Blackwell Publishers, 2010), 443-46.

of knowledge in economics had meant that the market could go on growing infinitely. Technology, in this model, would simultaneously make sense of the market and help speculators leverage it to their advantage.

New to network culture, the globalized residential real estate market led to cities and buildings that are largely empty, but rather act as sinks for global capital. Dubai is the obvious example. Venice, Italy is another: its population is largely composed of low-income service workers but the island is now dominated by pied-à-terres owned by the very rich who principally occupy them during the Biennale and Carnevale, forcing the service workers into housing on the mainland.[22] At night, the city is an empty ghost town, its buildings dark and empty. In global cities that still have an economic base, like New York, this logic extends to individual buildings and developments, leading to a phenomenon of individual buildings whose owners rarely visited them.[23]

The sentient city's promise of total information awareness and total connectivity for the technologically equipped is the urban embodiment of this financialized condition. Portable, wirelessly networked devices to interact with the city are an analogue to the high technology workstations in use on Wall Street. After all, the exchange floor has all but disappeared in the last decade. Virtually all exchanges have become purely electronic, existing in anonymous structures outside of the city where, presumably, they will be safe from terrorism. On the last great trading floor in the United States at the New York Stock Exchange (NYSE), traders have moved away from open outcry, embracing workstations that promise to deliver them total information awareness, prompting Vanity Fair to include Bloomberg Terminals on its list of 100 causes of the economic crisis since they led "Wall Street to believe it was invincible." Bloomberg Terminals, the writers of the Socializing Finance blog point out, not only give total information awareness, they contain an instant messaging service that helps establish market consensus and reinforce its internal logic.[24] The promise of the sentient city does the same for us, producing consensus that we are all players, ready to take advantage of the action, just like the traders once we have enough technology. But just as those traders are increasingly ephemeral — already some 75 percent of the trades on the NYSE are automated trades between algorithms set up to take advantage of millisecond-level fluctuations in the market — so, too the new city threatens that we will become ghosts, like the last inhabitants of Venice.

22. Wolfgang Scheppe et al., Migropolis: Venice: Atlas of a Global Situation (Ostfildern: Hatje Cantz, 2009).
23. Christine Haughney, "It's Lonely at the Plaza Hotel and ..." The New York Times (February 17, 2008) http://www.nytimes.com/2008/02/17/fashion/17plaza.html
24. Yuval Millo, Socializing Finance, "The Power of Market Devices on Vanity Fair," http://socfinance.wordpress.com/2009/09/05/the-power-of-market-devices-on-vanity-fair/

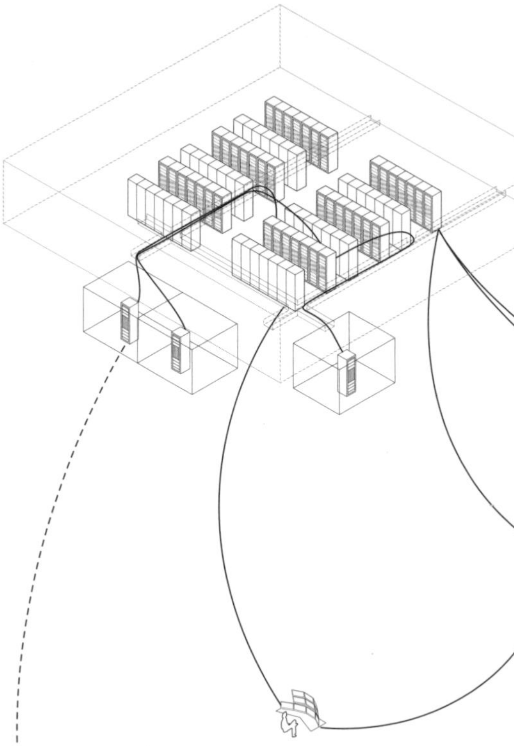

(3) Data Center
Mahwah, NJ

The contemporary stock exchange is an arena where financial competition far exceeds the limits of human performance or even perception. Since the 1970s, computerization has facilitated the flow of capital in financial markets; within the past few years, algorithmic and high-frequency trading practices have seized opportunities in the market that exist in micro-second windows. 40 to 70 percent of all equity trading on every stock market in the country is now performed by non-human agents and instruments: high-speed computers belonging to financial firms, co-located in the same facilities that house the machinery of the stock exchange, bridge the physical gaps of a spatial order in which financial information travels at the speed of light. Low latency, uncompromisable security, and high throughput (aggregated across tens of thousands of users) are imperative design criteria.

(4) Live-market Feeds
New York, NY

Two kinds of networks aggregate market data. The first, epitomized by the standalone Bloomberg Terminal and its portable version, delivers it securely in real time to traders and their clients. Low latency real-time ticker data streams, business wire news and other perishable data spur a perception of total informational awareness and reinforce market consensus. The second, more public system, epitomized by online presence of the Wall Street Journal and the Financial Times as well as free financial services offered by services like Yahoo! and Google, appeals to the everyday investor. With investment in financial instruments now naturalized through 401(k) accounts, these services reinforce the idea that everyone is a player and everyone can get rich, prompting investment that is then taken advantage of by financialization.

(2) E-trading Workstation
New York, NY

Electronic trading prompted the shift away from face-to-face trading to face-to-screen trading, contributing to the diminishing significance of a physical space for financial exchange. Even the floor of the New York Stock Exchange, a space associated with trading and one of the last remaining exchanges with a physical location on which trades transpire, is now dominated by workstations that display vast amounts of rapidly-changing data.

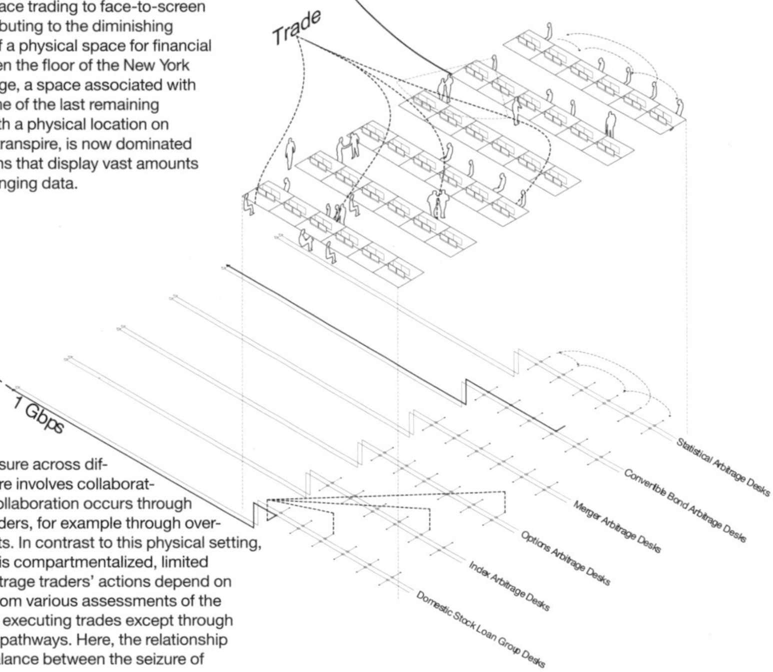

(1) Arbitrage Trading Floor
New York, NY

In arbitrage, a trade involves hedging exposure across different properties of evaluation, and therefore involves collaboration amongst different groups of traders. Collaboration occurs through impromptu social interactions between traders, for example through overheard conversations and tips about[1] markets. In contrast to this physical setting, the virtual architecture of investment firms is compartmentalized, limited in access for security purposes. While arbitrage traders' actions depend on the gleaning of associations and insights from various assessments of the markets, trader groups are precluded from executing trades except through their specialized applications and network pathways. Here, the relationship between actor and network hinges on a balance between the seizure of opportunity and control for volatility.

[1] Daniel Beunza and David Stark, "Tools of the Trade: The Socio-Technology of Arbitrage in a Wall Street Trading Room," Industrial and Corporate Change 13, no.2, (2004): 378.

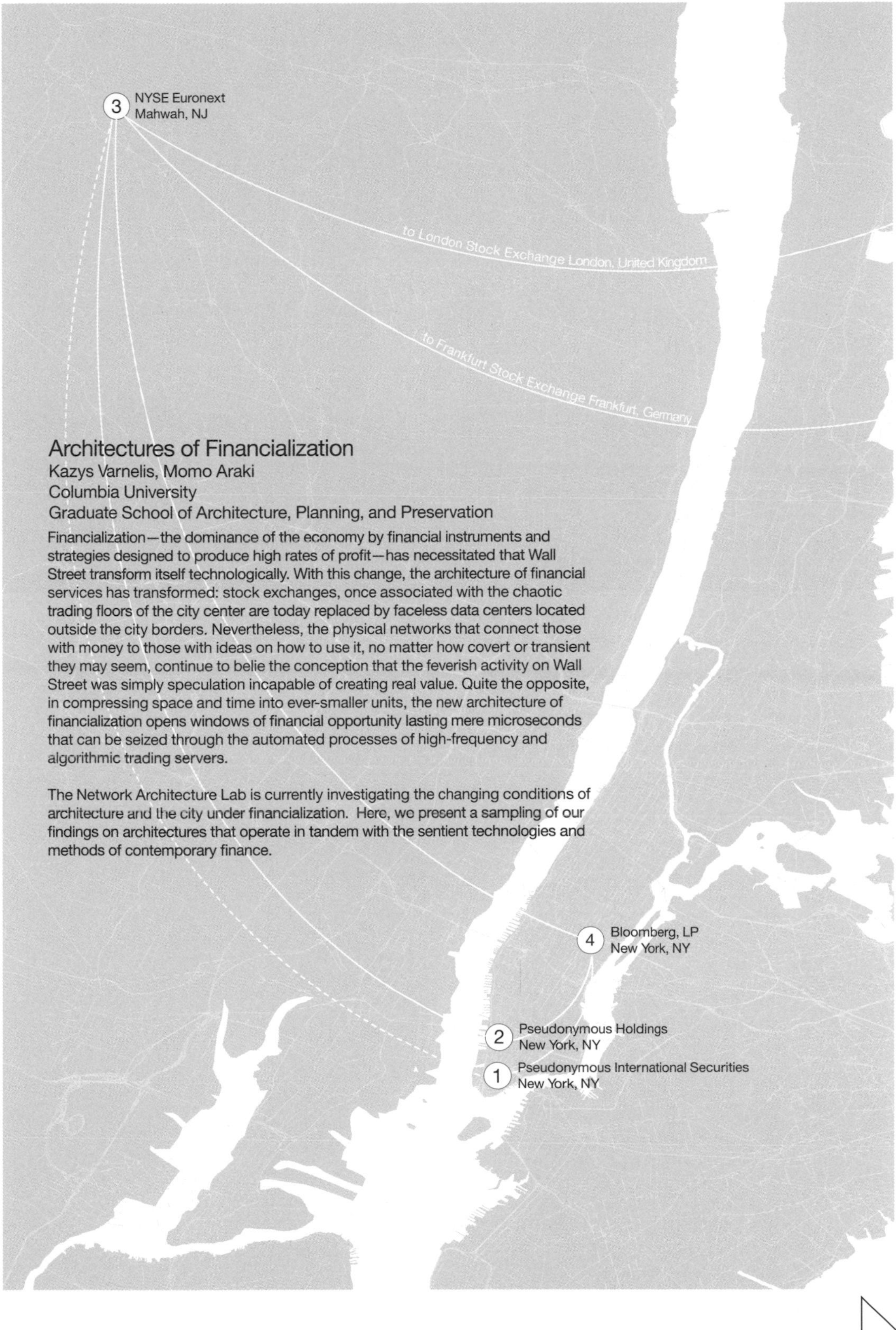

Architectures of Financialization

Kazys Varnelis, Momo Araki
Columbia University
Graduate School of Architecture, Planning, and Preservation

Financialization—the dominance of the economy by financial instruments and strategies designed to produce high rates of profit—has necessitated that Wall Street transform itself technologically. With this change, the architecture of financial services has transformed: stock exchanges, once associated with the chaotic trading floors of the city center are today replaced by faceless data centers located outside the city borders. Nevertheless, the physical networks that connect those with money to those with ideas on how to use it, no matter how covert or transient they may seem, continue to belie the conception that the feverish activity on Wall Street was simply speculation incapable of creating real value. Quite the opposite, in compressing space and time into ever-smaller units, the new architecture of financialization opens windows of financial opportunity lasting mere microseconds that can be seized through the automated processes of high-frequency and algorithmic trading servers.

The Network Architecture Lab is currently investigating the changing conditions of architecture and the city under financialization. Here, we present a sampling of our findings on architectures that operate in tandem with the sentient technologies and methods of contemporary finance.

Alex Rivera, Sleep Dealer (2008) Courtesy of Alex Rivera, www.alexrivera.com.

YOUR MOBILITY FOR SALE

TREBOR SCHOLZ

Mary, an attorney-at-law in Chicago, drives down the Dan Ryan Expressway. It's January 2007 and like every morning, she passes by a large digital billboard that visually commands the landscape. And what's more, it recognizes her, driving by in her Mini Cooper. A chip in the key fob of her car sends a signal to the billboard, which prompts a public message that could read something like "Mary, moving at the speed of justice." This promotional dispatch is shown next to a larger-than-life depiction of the car; it brightens her day.

As part of an advertising campaign, drivers of that particular vehicle who live in Chicago, San Francisco, Miami, or New York City could select whether or not they would like to see their name in big letters spelled all over the horizon. They could also stun fellow turnpike travelers with self-formulated pithy notes that would light up the moment they'd approach such billboards.

This PR campaign was popular in those four pilot cities, but some politicians and privacy advocates were not so amused. For them, the tiny, highly affordable radio frequency transponders used in this personalized advertising program are far more associated with colossal transnational corporations that manage the flow of containers and pallets of products. Privacy champions worry that the minuscule chip in Mary's key fob is not just a technological publicity stunt. In fact, it can provide valuable insights into her whereabouts and day-to-day behavior. For-profit organizations collect information from personalized location-aware advertisement and after it is analyzed, crosschecked, and visualized in multicolored info-graphics, it makes Mary's sentiments toward particular brands more predictable. The geospatial Web adds a new axis to Axel Brun's idea of the produsage; now, also movements through the city create content that has economic value. From the much more consequential perspective of governmental surveillance, Mary's behavioral patterns are registered and when she suddenly diverges from those points of reference, she becomes suspicious.

In the early days of the Internet, experts and novices thought of it as a space that was completely separate from actual life; and science-fiction author William Gibson introduced the term "cyberspace." But today it's clear that the Internet doesn't just converge with technologies of the screen, it is also entangled with wireless and mobile technologies, biometric surveillance devices, and linked to sensors and devices that are worn on the body. In this essay, I refer to this bundle of distinct technologies that rendezvous at the intersection of the Net and the built environment as The Internet of Things.

Today, objects imbued with artificial intelligence allow us to do more and more for ourselves and for others, but at the same time they also make us more and more vulnerable to being used: by each other, the US government, and the commercial sector. That is why it is paramount for architects, urban planners, artists, and the public at large to stay vigilant when it comes to the perils of the Internet of Things, which include potential breaches of geospatial privacy, "mission creep" in Fusion Centers, the often-unknowing expropriation of economic value from our day-to-day activities, and the intensifying "sanitation" of public access to the Internet through proprietary devices like the iPad.

The digital bait and switch economy is fueled by the booming data mining industry, which will become more proficient and sophisticated in the way it constructs and uses the data mosaics that stand in for us.

In this essay, I'm viewing these issues through the lens of what I call "geospatial labor." At the time of this writing, such digital labor is not an emotional hot button issue like the unprecedented oil spill in the Gulf of Mexico, the European song contest, or the World Cup. But over the next two or three years, the importance of geospatial labor will become unambiguous, provoking much moral indignation. Already now, more than 500 million people have wrapped their lives in the social networking service Facebook, which will soon roll out a geospatial location feature. Likewise let's not forget the nearly two billion Internet users and five billion mobile phone subscribers who will all be affected by the intensification of data collection and analysis.

The instruments with which these data are collected are largely hidden from view, they are pervasive, and they are changing so rapidly that it is basically impossible for anybody with a job and family to stay abreast this process.

Geospatial labor sounds awfully abstract, does it not? Let's make this more real by looking at some examples. The military has been using geo-tracking devices for many years in submarines and on airplanes. Sailors can be pinpointed and rescued. In the civilian market, the tracking of pets and livestock is the most common application thus far. Companies like Petsmobility.com and RFIDpet.com allow customers to introduce invisible fences and fewer cats end up in shelters. DigitalAngel.com makes seniors with Alzheimer's disease traceable. If I had a hundred-year-old grandfather who suffers from dementia I would be happy to pay Digital Angel to keep track of him on his urban wanderings. More widely known examples include EZ Pass, FasTrack, TxTag, and other systems that automatically collect your toll on highways or bridges across the United States. The EZ Pass, originally designed solely for this purpose, is now also used to produce evidence in contested divorce cases. [1]

Nike sneakers can now share your workout cycles and calorie loss with friends online, while also providing this information to the company that produces the running shoes. In the past, you may have been highly selective about the people for whom you removed your shirt, so that they might check your vitals. Today, with the help of Band-Aids that are embedded with circuitry, we can report our body temperature anonymously to the Web in real time. And if you really need to go to the hospital, don't forget to have a last look at your umbrella on the way out; it may light up when a cloudburst is imminent.

In the world of the Internet of Things, nothing will be free. Geospatial labor will be comprised of an-ever shifting topography of experience; it'll be invisible, fast, and everywhere. With PayPass, Mastercard hopes that the slight gesture of the hand that it takes to make a payment will become second nature.

1. E-ZPass records out cheaters in divorce court. August 10, 2007. http://www.msnbc.msn.com/id/20216302/ accessed July 25, 2010.

Or think of Mary on the Dan Ryan Expressway; her physical whereabouts become part of a commercial record.

What is so petrifying about being tracked? Isn't it a fair trade-off for the conveniences and the genuine utility that you derive from many network services. But do you actually know the stories that are told about you, to whom they are sold, and for what purpose? Have you considered the examples outlined above in their totality, rather than in isolation? Surveillance comes in many creeds: we watch each other, and we are tracked by governmental and commercial entities. The flows of data have changed in radically worrying ways.

Once a vision of a global data field — The Internet of Things— is enacted, it'll be hard to convince law-enforcement and government to refrain from using the collected data beyond the scope of its original objectives. Fusion Centers are a particularly lurid example. Late in the fall of 2001, this is what happened with Fusion Centers that were set up by the government to fight terrorism and detect the activities of foreign spies.[2] Today, there are over 70 such centers in the United States where an FBI agent may sit next to a Highway Patrol officer and the representative of a multinational corporation. Initially, the sole objectives of these centers were to merge information from government sources with data provided by the private sector to prevent the country from further terrorist attacks. Now, law enforcement would also like to use it to catch bank robbers and detect insurance fraud. A foreseeable next step is the use of Fusion Centers to identify political "troublemakers" at protests. There is, in fact, already one precedent for such abuse. [3]

Whatever your view of Fusion Centers is, notice that now it isn't only watchful eyes that follow us, it is also information gathered from sensors that gives an account of our real-time activities. We are all real-timers now and it is worth noting that far more people are hypersensitive about their geospatial privacy than about the public nature of their profiles on social networking sites. Whether we like it or not, we are exposed to sensors and the data extractors that analyze their output. It is worth stressing that the aggregated information is drawn from public sources but the analysis of aggregated data sets creates information about people that goes far beyond what was available in an individual database. In my view, it is also problematic that Fusion Centers provide for-profit organizations access to such enormous depth and scope of information.

Today, bandwidth, processing power, and storage have become inexpensive. This is why services such as FourSquare look and feel like a perpetual happy hour but everything in the digital economy must be paid for somehow. Free always comes at a price. Tomorrow's FourSquare user could be yesterday's no-collar worker, last week's white-collar worker, and last year's factory worker. Invariably, there are costs and each time, you pay a price. We are tenants on commercial real estate and our land-fee is paid for — almost unnoticeably — with our attention, data, and content.

Contrary to print, radio, and television, what generates economic value on the Internet is embedded in the medium itself, it

2. ACLU. What's Wrong With Fusion Centers. December 5, 2007. http://www.aclu.org/technology-and-liberty/whats-wrong-fusion-centers-executive-summary accessed July 25, 2010.

3. While a detailed description of the case of Ken Krayeske, a law student in Connecticut one time Green party campaign director who had been arrested but not convicted at an antiwar rally. In January 2007, Krayeske stated his opposition to Connecticut Gov. Judy Rall and through a blog post he invited people to protest at Rall's inaugural parade. Law enforcement received a threat list, which was compiled with the help of the local Fusion center, and Krayeske was promptly arrested when he appeared at the parade and started to take photos.
Ken Krayeske fusion center interview. http://www.youtube.com/watch?v=_ZKTNJCII78&feature=related (accessed July 25, 2010).

dissolves into the background. Labor doesn't look like labor at all and as Paolo Virno says, life itself is put to work.

The paradox of geospatial labor is that the means of communication are in the hands of a multitude of Internet users while only a handful of companies are making lots of money from what looks like a free service. These companies have control and they gain economic value disproportionately to those people who are using a given online service. This is why it still makes sense to talk about labor in the context of the network society.

Online, you publish a website and thereby provide information to Google. Every one of your clicks contributes to their core business. Technology publisher Tim O'Reilly articulated this bluntly: "... they are participating without thinking that they participated. That's where the power comes."[4] The power that O'Reilly is talking about is the power of corporations. It is not the power of users who unintentionally participate. It is true, Google's half million servers and almost 40 data centers are costly. It doesn't come as a surprise that Google wants to become an electricity provider because electricity is one of their main expenses. And Google also has to pay for developers' bandwidth and much more. Sure, it can be a bruising business, but Google's net worth is $20 billion, and the company earned $4 billion in profits in 2008 alone. Profit from digital labor is real.

In the past, the introduction of new technologies has been slower, allowing for the development of countermeasures to invasions of privacy. For example, take the introduction of surveillance cameras near cash machines, which led to compulsory signage alerting passersby that they are being recorded. Or, think of the way Caller ID led to mechanisms that allowed certain callers to block it.

With Google or Facebook, new features are rolled out with lightning speed. Seriously, could anybody really have enough time to dedicate themselves to the 50 clicks that it takes to set Facebook's privacy settings in their favor? Surely, Facebook changes its policies constantly and what was once 50 clicks may now be five (and then again 50 by next week) but being aware of that still requires an enormous amount of attention that only a digital elite is capable of. Similarly, look back to the uproar that was caused by the introduction of "cookies" about a decade ago. Today, many websites can't even be accessed unless cookies are enabled on the web browser, and few people seem to care. Overall, attitudes have migrated to the side of accepting a digital lifestyle that is public by default.

Now imagine these issues in the context of the Internet of Things when you don't even have a User License Agreement in front of you; you merely walk into a participatory environment and your geospatial footprint becomes somebody else's business. Jeremy Rifkin makes the assertion that friendship, love, and job, are inexorably being sucked into commercial life. Starting in the 1980s, when researchers began to embed network censors into the built environment, ubiquitous computing has turned the city into a site of value production. Urban studies scholar Laura Y. Liu even uses the term Sweatshop City.

4. O'Reilly, Tim. In Lessig, Lawrence. Remix making art and commerce thrive in the hybrid economy. New York: Penguin P, 2008. Print. 224.

In the film The Creators of Shopping Worlds (2001), Harun Farocki, demonstrates how surveillance technicians, mall owners, "bread display architects," and planners study the movement and gaze of shoppers in detail. Who looks up or down, and when? There's a whole industry behind the question of manipulating people into making purchases and every year some 6000 experts meet in Las Vegas to compare notes. In the world of ubi comp, we are all test subjects and information dynamos in the petri dish of emerging forms of commercial and governmental surveillance.

In the Sweatshop City, our data body never sleeps; it is always already present, from sensors in book and music stores that know which novel we checked out in the library, or which artists' music we recently purchased on iTunes, to those in the mall that know with whom we are in a serious relationship, as well as significant dates, like anniversaries and birthdays, when purchasing potential is higher. It is important to understand that these all-pervasive recording systems output data that is of real consequence to our lives. While most Americans have never heard of data aggregation companies like ChoicePoint, they directly impact their lives. ChoicePoint plays [5] a role in the hiring process of millions of Americans. The data that this company aggregates is consulted when you apply for a loan, want to rent an apartment, or apply for health insurance. Citizens do not have access to the data that are collected about them and have no way of correcting false information.

With the Internet of Things, we will not just be "glass consumers" whose purchasing power can be directed, rather absolutely all aspects of our lives will have a deep record. It's unavoidable to remember the dark visions of the Total Information Awareness Act put forward by former National Security Adviser Admiral John Poindexter. With the help of the Internet of Things, Poindexter's plan to create one giant meta-database with files about each one of us would make 1984, J. Edgar Hoover and even the smell archives of the former East German Stasi look (and smell) like an old shoe.

Even Secretary of State Hillary Clinton closely associated security with technology in her January 2010 policy speech on Internet Freedom. [6] Sitting in the audience that day, I was surprised when she offered a tiered approach to online anonymity which she deemed to be good when it helps political dissidents in Iran for example, but which she opposed when it helps the enemies of the United States. With Internet protocol IPv6 such distinctions will not be possible: it will make it easier to identify anyone online, regardless of whether or not he/she is a law-abiding citizen or a terrorist. Furthermore, despite its significant achievements, the current administration has re-authorized three contentious provisions of the Patriot Act. Since 2004, the US government started over 200 data mining programs, over 35 of which link specific data to individuals. In 2006, Eric Lichtenblau of the New York Times revealed the fact that the National Security Agency recorded millions of phone calls by US citizens on domestic soil. [7] Today the "Terrorism" Information Awareness Act [TIA], provides the government with legal justifications to "ingest" information from a score of databases,

5. ChoicePoint. http://atxp.choicepoint.com/ (accessed July 25, 2010).
6. Foreign Policy. Internet Freedom. The prepared text of U.S. of Secretary of State Hillary Rodham Clinton's speech, delivered at the Newseum in Washington, D.C. <http://www.foreignpolicy.com/articles/2010/01/21/internet_freedom?page=full> (accessed July 25, 2010).
7. Lichtblau, Eric, Risen, James. 2005. Bush Lets U.S. Spy on Callers Without Courts. The New York Times, December 16. http://www.nytimes.com/2005/12/16/politics/16program.html?_r=1

blogs, e-mail traffic, intelligence reports, and other sources including government documents to create data portraits. Interestingly, the government relies on private data brokers to provide the desired information, in many cases in conflict with the 4th amendment. Again, what would this scenario look and feel like, fueled by some incarnation of an Internet of Things?

But surveillance is not the only concern. The customizable location-based spectacle may also lead to more choices; doors may open that we didn't even know existed and every step we take, deeper into the tissue of the city, will confront us with more choices. Barry Schwartz, the author of the The Paradox of Choice: Why More Is Less, warns us of a vertigo of trivial choices, which he thinks will make us more anxious or even depressed. He writes: "though modern Americans have more choice than any group of people ever has before, and thus, presumably, more freedom and autonomy, we don't seem to be benefiting from it psychologically." [8] We may feel like our life is more and more rushed and complicated, we are threatened by tyrannical waste of time.

There is no room in this essay to go into detail about the various kinds of digital labor or to really discuss the increasingly difficult distinctions between work and non-work, the playground and the factory, public and private, users and operators, exploitation and expropriation. Many of these categories have been blurred, intermingled, or used interchangeably. Despite this conceptual murkiness, it is worth noting, however, that I'm considering digital labor merely as a shift of traditional labor markets to the Internet. There is really nothing completely new, but the forms of labor have changed; some accounts of 21st century labor are different from those of 20th century labor. From the fast food restaurant to the grocery store and airport, small acts of labor have been taken over by the consumer. In the 18th century, Diderot and d'Alembert celebrated the virtues of labor in their Encyclopedie. Surely, today they would have to chronicle additional forms of labor including the voluntary writing of book reviews on Amazon.com, and the unpaid customer service that people perform on Verizon's and Apple's websites by responding to technical questions on support forums. They would also have to think about the paid, but uninsured, work facilitated through services like Mechanical Turk [9], LiveOps [10], or Txteagle [11]. The latter, hands out work tasks to people in sub-Saharan Africa through their cell phones. Just like in Alex Rivera's riveting film Sleep Dealer, these distributed labor practices allow access to a labor force without the employer having to deal with the needs, contentions, and rights of the workers. They get all the work without the worker, as Lisa Nakamura analyzed.

This essay started with a story about an attorney-at-law in Chicago. In her particular case, the company behind the ad campaign claims that they made no use of the data gathered. This may be so but for all I know there are many companies who wouldn't hesitate to monetize collected data, not for a second. With the Internet of Things, the digital economy has reached the street and we need to map and theorize this situation, but more importantly, we need to think about tangible solutions.

8. Barry Schwartz, The Paradox of Choice: Why More is Less (New York: Harper Parennial, 2004), 99.
9. Amazon.com's Mechanical Turk. https://www.mturk.com/mturk/welcome (accessed July 25, 2010).
10. Liveops. http://www.liveops.com/ (accessed July 25, 2010).
11. Txteagle. http://txteagle.com/ (accessed July 25, 2010).

Ted Nelson's idea to run the whole Internet as an exchange system of micro-payments for contributions to the network: it simply isn't feasible. Digital labor and the range of underlying technologies are characterized by data lock, pervasiveness, and the invisibility and constant change of the tools of surveillance. What can we ask in return for this violence of participation? What can we demand in return for unwillingly and unknowingly providing profile data, social graphs, and activity streams to for-profit organizations?

First, nobody should lose ownership of their body; our data body, the stories that are told about us, need to be accessible to us so that we can check if they are accurate. Who sells these stories to whom and for what purpose? How can we opt-out of generating such narratives? Which mechanisms could be put in place to allow us to select which particular information we would like to share with a particular party? Crucially, these records must be accessible to us without much effort; if it takes thousands of clicks to reach such security reports, it becomes too arduous for us to follow up. Multiple versions of a Bill of Rights for users of social media have been circulating for several years now. One version on opensocialweb.org culminates with the demand that user data should be portable, accessible, and verifiable. Such manifestos are important because they insist that users have a right to have rights and that they should in fact compel companies to make transparency and access to our data a point of competitive advantage.

Throughout this essay, described in broad brush strokes, I have set apart the benefits from the threats of an Internet of Things. It would behoove many policymakers, entrepreneurs, and those who overzealously exercise governmental power to use the European Union as a reference point. Here, governments take a much harder stand when it comes to defending the rights of their citizens online. But no matter in which country one lives, the Internet of Things will bring about rich experiences for many of us in the overdeveloped world, but it will also make a very small number of people unjustly wealthy.

COMFORTS, CRISIS, AND THE RISE OF DIY URBANISM

MIMI ZEIGER

Over the course of its 138-year history, Popular Science Monthly has developed a tendency to feature illustrations of rocket ships, ecotopias, and fantastic vehicles on its cover. Finding a photograph (however posed) among these cover images — some more akin to the comic books of the time than current events — is a rare occurrence. But in May, 1941, two soldiers, more cherubic than fierce under First Infantry Division helmets, graced the cover of Popular Science Monthly. A 60-mm mortar, the true pin-up of the composition, splits the pair. The image, the date (just months before Pearl Harbor) and the tagline, "Our Infantry's New Weapons, Page 64," combine to broadcast a nation's trust in technology.

However unusual, the cover is just the surface of an issue full of tips and how-tos for a country on the brink of crisis. Page 67 offers a photographic breakdown of "New Guns and the Men Behind Them." And page 70 paints a picture of modern warfare in contrast to Great War trenches: "It is warfare in the open, in the field, in woods, in villages; quick, short thrusts, and when the final surge forward comes Dick is virtually on his own, and therefore he must be trained not only in discipline, but in initiative," writes John Watson with a certain flourish. [1] To linger on Watson's sentence, to break it apart clause by clause, is to not only track the development of a new kind of "open" combat that takes the built environment as its theater, but also the development of a solider ready to think on his feet, to "virtually on his own" blaze an independent path, to do it himself.

But to stop and mull here is to miss a table of contents rich with stories prescient of our own contemporary condition: "Protecting Homes for Air Attack," "Are Skyscrapers Bombproof?" and "Fly Yourself in a Rented Plane." [2] Hidden behind a title less provocative than others on the line-up, "Laboratory for Young Scientists" announces the establishment of the American Institute Laboratory at 310 Fifth Avenue in New York City. The address is important. It's IBM's headquarters. The company volunteered two floors of choice real estate to host budding scientists. Westinghouse Electric & Manufacturing Company filled those offices with top notch equipment — drafting boards, radio sets, glass blowing tools, and drill presses.

According to the article, American Institute of the City of New York founded its "junior branch" in 1928 explicitly to "direct and utilize the imaginative faculties of youth."[3] And at the time of writing, American Institute Laboratory was open to the 30,000 science club members across the United States. Each year budding scientists and experimenters would gather at the Museum of Natural History for a science fair. The competitive fair, buoyed by corporate and civic sponsorship, rose in popularity over the next few decade, it's mission dovetailing with the need for a new kind of GI during World War II, then as post-war suburban kitchens filled with appliances and ever more complex chemical solutions, science fairs served a double purpose. They trained American youth and offered the public a perceived transparency to scientific and technological developments of the time. If a high-schooler could build a crystal radio set, the country's future is in good hands.

1. John Watson, "Infantry's New Weapons Give Fire Power and Punch", Popular Science Monthly, May, 1941, pgs. 64-70. http://www.popsci.com
2. Popular Science Monthly, May, 1941, pg. 2. http://www.popsci.com
3. "Industry Gives a Laboratory to America's Young Scientists", Popular Science Monthly, May, 1941, pgs 49-51. http://www.popsci.com

POPULAR SCIENCE *Monthly*

INDUSTRY GIVES A LABORATORY TO AMERICA'S YOUNG SCIENTISTS

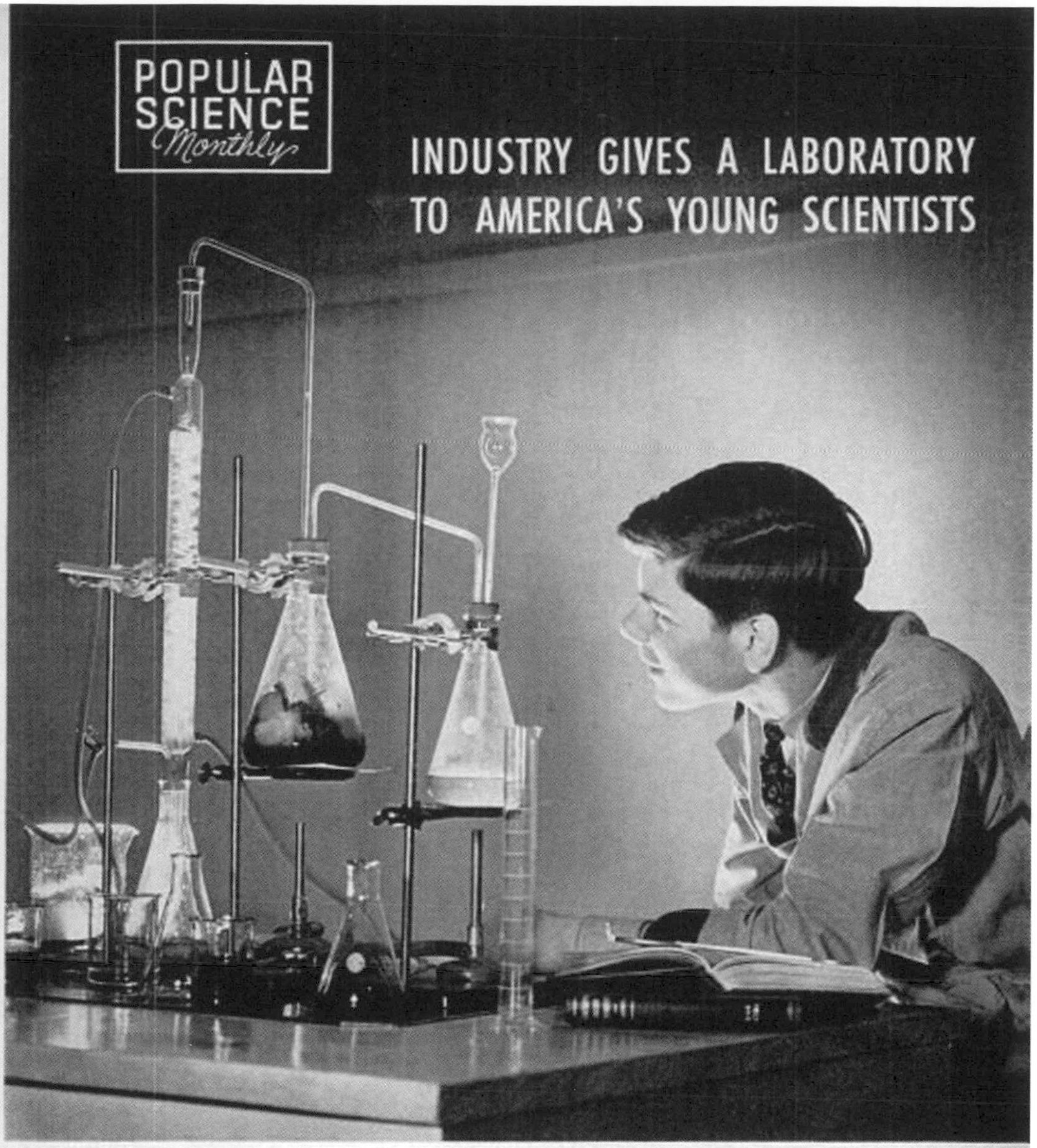

YOUTHFUL IMAGINATION, an inexhaustible national resource, is being developed along scientific lines by the American Institute of the City of New York. This organization, chartered in 1828 and devoted throughout its existence to the promulgation of science and the encouragement of American industry, established its junior branch in 1928 and recently has intensified its efforts in this direction through the American Institute Laboratory at 310 Fifth Avenue, New York.

Its aim is to direct and utilize the imaginative faculties of youth which, since the founding of the institute, have been turning more and more toward science and mechanics. Under its wing are more than 730 juvenile science clubs, scattered throughout the United States, its possessions, and foreign countries. Some meet in high schools, some in settlement houses, and some are spontaneous youthful organizations with cellar or attic laboratories and club rooms. In the aggregate there are

Popular Science Monthly, May 1941.

Home Ec

The legacy of Popular Science and science fairs carries over into contemporary culture.

In Made by Hand: Searching for Meaning in a Throwaway World, author Mark Frauenfelder, founder of the geeky and influential blog, Boing Boing, and editor in chief of Make magazine, connects the rise of current DIY movements with network culture. As to the garage myth of Make, he writes:

We looked at do-it-yourself magazines from the past — the 1940s through 1960s were a heyday for do-it-yourselfers. Popular Science contained quite a few articles devoted to making things such as go karts from lawn mower engines and modernistic coffee tables from plywood and ceramic tile.

Dale [Dougherty] and I noticed that these kinds of projects, along with traditional crafting and homesteading activities, such as gardening, raising chickens, keeping bees, and preserving food, had again become popular, due in large part to the great information-delivery capability of the Web. People were rediscovering the joy of DIY. We decided that the magazine should be a celebration of the kind of making, experimenting, and tinkering that was being chronicled by enthusiasts on the Web.[4]

Frauenfelder's observations, subsequently successful magazine, and search for meaning memoir sketch out a DIY history rooted in the domestic arts. He spends an entire chapter narrating the destruction of his front lawn in order to plant a vegetable garden along the lines of artist Fritz Haeg's *Edible Estates*. Another chapter charts the trials and tribulations of setting up a bee colony at another suburban home in Tarzana, California. (Apparently, locally grown honey is the modern-day homesteader's mark of authenticity — a blatant rebellion to the kind of honey sold in bear-shaped vessels in chain supermarkets.)

Yet, even as it is celebrated as activist, the domestication of DIY is a nostalgic act made possible first by pervasive home computing and now by the seamless integration of handhelds and tablets into everyday life. Picture the digital DIY still life: homemade apple pie and iPad. Just as mid-century science fairs' optimism was colored by atomic fear — a threat that negated public space opting for the comforts of home — the kind of DIY professed by Make (as well as other contemporary magazines such as Martha Stewart Living or the hipper younger sisters, Readymade and Craft) is clouded by terrorist threats. In the weeks and months after 9/11, "nesting" cropped up as a feel good term to describe why everyone wanted to stay home and sit on the couch eating artisanal macaroni and cheese. Self-reliance translates to an almost smug delight in the authentic and the handmade.

The Hack

"I was charmed by a perspective of the world as a hackable platform, something to be remade and remodeled to his exacting, eccentric, yet infectiously appealing aesthetic sensibilities,"[5] writes Frauenfelder, describing his first encounter with Make tool reviewer Mister Jalopy. Leaving behind the aesthetic

4. Mark Frauenfelder, In Made by Hand: Searching for Meaning in a Throwaway World, Penguin Group, New York, 2010. pg 15.
5. Ibid, pg. 21.

sensibilities (which seems to by default fall back into lifestyle choice), let's focus on the hack. Make magazine spawned the Maker Faire in 2006. Self-consciously geeky, the faire celebrates the constructs of tinkerers, hobbyists, crafters, and artists. Makers crack open objects that define our everyday: cell phones, digital cameras, or computer hardware. Not only do they produce a willful transparency into the stuff of our post-industrial existence, Makers connect together online and in real life, forging networks of DIYers unfazed by changing technologies.

When Frauenfelder's co-founder, Dale Dougherty, spoke at the Aspen Ideas Festival in July 2010, he extended the individuated DIY ethos into what he called a "hands-on imperative" or, the idea that aggregated efforts of makers could add up to something that changes the world—science, education, and business. "I'd like you to think about the hands-on imperative as not being limited to things that you can grab with your hands," Dougherty, as quoted in an online piece in The Atlantic. "Today, people think about getting their hands on data, getting their hands on their DNA."[6] There's potential to expand Dougherty's call to make into a kind of DIY urbanism. The hack can apply both metaphorically and literally to the built environment, so why not use the tools of ubiquitous computing to take DIY out of domestic enclaves and into public space? As an extension of the kind of open source methodology found in software development, Matthew Fuller and Usman Haque with their Urban Versioning System 1.0 outlined a possible tactical framework.[7] Although Fuller and Haque set up a series of constraints, each of these restrictions harbor DIY's mutable and ad hoc sensibility from informal settlements to self-build architecture.

6. Alex Madrigal, "Can Building Robots Reboot Education?", The Atlantic, July, 9, 2010 http://www.theatlantic.com/special-report/ideas/archive/2010/07/can-learning-how-to-build-robots-reboot-education/59488/
7. Matthew Fuller and Usman Haque, Situated Technologies Pamphlets 2: Urban Versioning System 1.0, The Architectural League of New York, New York, 2010

PARK(ing) Day. Photo Michigan Municipal League.

However, Mathew Passmore co-founder of the San Francisco-based design studio Rebar is also taking DIY into broader territory. Rebar is best known for developing PARK(ing) Day, an exercise in participatory urbanism in which parking spaces are transformed into temporary parks. He cites the work of experimental art and architecture group Ant Farm (founded 1968 by Chip Lord and Doug Michels) as precedent to the current DIY movement.[8] Passmore is taken with the group's mobile and inflatable structures used for events and happenings (a legacy reflected in Rebar's work), but Ant Farm's relationship to media is just as important in building a bridge to contemporary practice. Consider the piece *Media Burn*, in which on July 4, 1975 two drivers dressed as astronauts pilot a modified Cadillac through

8. Matthew Passmore, "Participatory Urbanism: Taking Action by Taking Space", SPUR Urbanist, February 2010.

Ant Farm. Interior view of the Phantom Dream Car, from Media Burn, July 4, 1975; performance, Cow Palace, San Francisco. Photo: Curtis Schreier. University of California, Berkeley Art Museum and Pacific Film Archive. Purchase made possible through a bequest from Thérèse Bonney, by exchange, a partial gift of Chip Lord and Curtis Schreier, and gifts from an anonymous donor and Harrison Fraker.

The Artist-President (Doug Hall) addressing the crowd at Media Burn, July 4, 1975. Photo: Chip Lord. University of California, Berkeley Art Museum and Pacific Film Archive. Purchase made possible through a bequest from Thérèse Bonney, by exchange, a partial gift of Chip Lord and Curtis Schreier, and gifts from an anonymous donor and Harrison Fraker.

a flaming pyramid of television sets. While the crash through the TVs is clearly the highlight of the event, a singular, smashing hack, it has to be noted that a video monitor placed between the car's bucket seats guided the drivers—an appropriation and critique of a technology soon to become ubiquitous. It is also impossible not to add Archigram's Instant Cities (1968) to the list of historical references, especially since the design proposal not only combines pop graphics (drawn equally from science, fantasy, and advertising) with media savvy optimism, but also because it comes with a darker, almost militaristic deployment.

Sensing Sensibilities

"Today, as computing leaves the desktop and spills out onto the sidewalks, streets and public spaces of the city, we increasingly find information processing capacity embedded within and distributed throughout the material fabric of everyday urban space," wrote Mark Shepard in his curatorial statement for Toward the Sentient City.[9]

Shepard outlines a smart city, an empowered city, an urbanism wired to react. However, he is quick to point out that the embedded technology is not neutral. It senses, gauging its behavior on our actions. He continues, "Imbued with the capacity to remember, correlate and anticipate, this near-future 'sentient' city is envisioned as being capable of reflexively monitoring its environment and our behavior within it, becoming an active agent in the organization of everyday life in urban public space." As much as the city is sentient, it is also hackable and prone to intervention. Its mutability gives rise to projects that can be categorized as DIY urbanism. These localized projects, such as *Amphibious Architecture* or *Natural Fuse*, counter the big crises of our own time — terrorism, catastrophic climate change, economic instability — with a transparent view into everyday infrastructures.

For example, *Amphibious Architecture*, a collaboration between The Living and Natalie Jeremijenko, dunks an array of interactive tubes in the East River and the Bronx River. Equipped with sensors, the floating buoys collect water quality readings and monitor for fish. Viewers text the fish a watery "What up?" and the fish report back on the state of their ecosystem. The architecture here is not just amphibious, but anthropomorphic; I sent the fish a message and the fish wrote back "Underwater, it is now loud," signaling the presence of marine life, and urged me to text friends AhoyAnchovie and HeyHerring.

This call and response between participants and fish is welcoming in its casual interface, with clever phrasing acting as a palliative for the uncertain condition of New York City's waterways and the ever present specter, the global state of water and the environment. Both participatory and self-built, the project could fall under the heading of DIY urbanism. Additionally, the science and gadgetry isn't hidden, but blatantly exposed in the methodical, how-to style of Popular Science. Photographs of the construction and installation reveal a fascination with capturing the components: circuit boards, cabling, LEDs, and sensors. (Interestingly, *Amphibious Architecture* was assembled at Studio-X New York, a workshop and gallery sponsored by

9. Mark Shepard, Toward A Sentient City Curatorial Statement, 2009. http://www.sentientcity.net/exhibit/?p=3

Columbia University's Graduate School of Architecture, Art and Planning that labels itself a collection of laboratories and trades in architectural experiments.)

Unlike homespun DIY, when the complex networks that provide the underpinnings of daily life are made visible, the results are not always comforting. Science fair-like in its low-fi display, *Natural Fuse* by Usman Haque, Dot Samsen, Ai Hasegawa, Cesar Harada, and Barbara Jasinowicz, comes across in the gallery as a cheerful, hands-on learning station. There is perhaps nothing less threatening than a houseplant. A planter, watered via tubing connected to recycled water bottles, is wired to a desk lamp. Designed to call attention to carbon offsets, a *Spathiphyllum* is hooked up to a device that tracks its CO_2 output. That local system is then tied into a citywide network of other houseplants, which regulates energy usage and carbon absorption. (Google Maps is used to geo-locate other plant sitters.) Sounds friendly enough, the plants thrive when everyone in the network gets along. But, if not "then the network starts to kill plants, thus diminishing the network's electricity capacity." Users have the option of being "selfless" or "greedy." It's a grim opposition. Plants live or plants die on collective decision.

You And You

The "collective" problemitizes the "you" in Do It Yourself and points to a broader understanding of DIY urbanism — a Cognitive Surplus [10] approach to infrastructure and public space. Weeels (we plus wheels), for example, uses iPhone apps and SMS tools to create an ad-hoc mass transportation service. Part livery cab, part ride-share program, multiple passengers are matched with drivers around the city. Created by David Mahfouda, founder of the Fixers' Collective, an old-fashioned, recessionary "social experiment in improvisational fixing and mending," Weeels has the utopian potential to be an efficient third-space between taxis and subways — an efficient and environmentally responsible way to connect underserved urban spaces in New York City. [11, 12]

Indeed, Weeels offers an fairly elegant solution to a host of complex urban issues: mass transit cutbacks, fuel conservation, neighborhoods in Queens or Brooklyn not reached by public transit and for reasons of class, race, and remoteness, simply not served by taxis. But even good intentions come with specters. A question on the FAQ page asks: "Is it safe to get into a car with a stranger?"

It is a query that tracks back to the beginnings of automotive history and is just as relevant now. There are underlying fears. As much as accessible technology brings people together in common cause, the reshaping of the city through DIY urbanism comes with dread that no science project, fix, or hack can easily alleviate. The answer given by Weeels applies largely to both the potential and limits for these kinds of projects: "...you can always choose to cancel the search or ride alone."[13]

10. Clay Shirky, Cognitive Surplus: Creativity and Generosity in a Connected Age, Penguin Press, New York, 2010.
11. http://www.weeels.org
12. http://fixerscollective.org
13. http://www.weeels.org/faq.html#is-it-safe-to-get-into-a-car-with-a-stranger

TOWARD THE SENTIENT CITY: EXPECTING THE EXTENSIBLE AND TRANSMISSIBLE CITY

ANNE GALLOWAY

As a social anthropologist and design researcher my work first seeks to describe the world and, through that description, to forge an understanding of it. Often this means that I am more interested in understanding how things come to be than I am in imagining what I think they could be, or prescribing what I think they should be. I believe that focussing on processes of becoming puts me right in the thick of the making and doing of everyday life, and by extension offers glimpses of how things can be undone or made differently. It forces me to identify who and what are at play and it allows me to ask how and why some things are actualized while others remain potentials, and yet others fade from the realms of possibility or plausibility. Bringing these interests to bear on the case studies presented in this volume involves asking one deceptively simple question: What current expectations stand to shape the "sentient city" of the future?

In studying the production of scientific knowledge, sociologist Michel Callon has outlined a set of social and material processes that can help us understand how sentient cities might emerge, and open up space to consider the roles that hopes, expectations and promises play in processes of emergence. First is *problematization*, or how ideas and things become indispensable; second, *interessement*, or how allies are locked into place; third, *enrollment*, or how roles are defined and coordinated; and fourth, *mobilization*, or how issues are represented to others. Not all *problematizations* result in *enrollment*, but if the *interessement* is successful then the actors move to define, coordinate and enroll themselves and each other into particular roles. Negotiations that take place during *problematization*, *interessement* and *enrollment* invariably involve more individuals than a given assemblage claims to, and indeed is able to, represent. This question of representation, or who speaks on behalf of whom, is of clear social, political and ethical concern, and if 'spokesmen' (i.e. people, things and ideas) are designated by putting "intermediaries and equivalences" into place, then looking at these things also allows us to see who and what are silenced or denied a place on the playing field. Continuing negotiations between the representatives seek to mobilize and commit absent or silent actors, and if the mobilization is successful then these relations will be accepted as 'real' and sometimes even 'normal.' Sociologist Bruno Latour calls these mobilized and assembled realities "actor-networks," and actor-network theory is more properly a methodology for tracing the associations that form actor-networks.

Using the concepts described above, we can see that non-linear movements and changes in trajectory, as well as path-dependencies and obduracies, are integral to the kinds of associations that might create sentient cities. These processes of recruitment, representation and association are also connected to expectations, hopes and promises. To expect something is to look for, or look forward to, something in the future; but expectations are also powerful performative and generative social phenomena in the present. A sociology of expectations looks to the affective roles of imagination and desire (i.e. the capacity to be moved) in shaping technological and social change. For example, expectations serve to attract interest, secure investment, establish relations, define roles and guide activities. In a world of problem-focussed design explorations, as well as within critical and creative research activities, expectations and hopes work like promises or binding agendas and obligations. Expectations can be positive or negative, and especially in the case of techno-science, are often put in terms of utopian or dystopian futures. Expectations in such cases are also associated with the belief that technoscientific progress is both a requirement and a promise, where practitioners assume a certain technological inevitability and feel obligated to deliver the best possible product, service or alternative solution in response. However, as we proceed along our dedicated paths, we also tend to ignore or exclude things that do not reinforce our assumptions or fit within our plans. And yet it is useful to note that expectations — like actor networks — are limiting rather than determining forces. Not all promises are kept, not all hopes get fulfilled, and not all expectations come true.

Urban computing and locative media applications — like the projects showcased in this volume — can be distinguished by their desire to *activate* urban spaces and social relations. Cultural geographer Mike Crang has written about "transmissible cities," or complex and hybrid informational spaces, and sociologist Mike Michael has described this complex sociality and spatiality in terms of endless disclosure, where new technologies work to open up rather than enclose places, and to reveal rather than conceal activities. In these scenarios, emergent city spaces and social behaviors are expected to become more affective and expressive, as well as more meaningful because they can create a sense of shared reality or co-presence. Relatedly, social knowledge becomes a matter of knowing interiority, or what is going on *inside of* or *under* the surface of things. In this vision of the city as an interaction design space, computation also extends *over* the built environment, transducting another layer of exteriority to be experienced. This extensibility and transmissibility of the city, along with an increased ability to be embedded within it, is a core expectation and promise shared amongst the cases here.

In terms of social, spatial and technological expectations, the least radical project is Anthony Townsend and team's Breakout!, which provides "tools to help office workers escape dull cubicles and conference rooms and re-locate their work in public settings." Working with familiar technologies like wireless Internet connectivity and social software, the project seeks to "explore the potential" of public spaces and transit by re-appropriating them as workspaces. First, this assumes that office work is dull and that people want or need to escape it. Second, it implies that the full potential of public space has not yet been realized and it assumes that people want "the entire city" to become their office. While I suspect that all sorts might agree with the first assumption, what seems to be missing from this scenario is the possibility that other people may not want public spaces to be colonized by office-work. The promise inherent in the project is a city where technological penetration of everyday life and the extension of work into leisure time are highly normalized. While we could argue that is already a promise fulfilled, we might also acknowledge that the temporary nature of the festival scenario offers respite from such a technologically totalizing vision of everyday urban life.

Taking the concept of networked devices in a more explicitly critical direction, Usman Haque and team's Natural Fuse project creatively addresses issues of urban electricity production and consumption. In this case, the project expects a future where people's desire to collect and transmit information will not wane, but we may not be able to produce sufficient power to feed our desires. By creating a networked infrastructure of "electronically-assisted plants that act both as energy providers and as circuit breakers," their proposal obliges us to consider the short- and long-term implications of actually making a sentient city run. Perhaps most interestingly, by extending the concept of a network well into the natural or organic realm, the project demonstrates the need for continued research and design in the area of "carbon-sinking" if we expect it to provide a workable solution to our energy needs. As the project description points out, "You might need 420 plants to offset your 50W lightbulb!" and that sort of practical reality suggests that continuing to develop and implement new technologies without consideration of how they will be powered, or how we might need to modify our consumption patterns, is a dangerous path by which to proceed.

Both the Living Architecture Lab and Natalie Jeremijenko's Amphibious Architecture and the SENSEable City Laboratory's Trash Track projects can be seen to critically and creatively explore aspects of urban transmissibility and extensibility by making the invisible visible. Like Natural Fuse, the Trash Track project uses networked technologies to explore and support critical infrastructure. By electronically tracing items through the city's waste management system, the project produces real-time visualizations of the journeys taken by things we throw away. In this case, the expectation or hope is that this knowledge will provide a richer understanding of how a city works, and encourage more bottom-up approaches to its sustainable management. However, if this hope or expectation creates a path-dependency for future research and design, we risk overlooking or ignoring the possibility that this desire for informational awareness is not shared, or that it may not result in the anticipated citizen actions. Similarly, the Amphibious Architecture project comprises an underwater sensor network that monitors "water quality, presence of fish, and human interest in the river ecosystem." However, rather than creating online visualizations of normally unavailable or inaccessible information, the project includes above water light displays that encourage on-site interaction by allowing people to access the sensor information via text message. An assumption here is that being able to transmit data across physical barriers enriches people's experience of the city and its natural environment, and Amphibious Architecture further creates an expectation that people will both want and value these kinds of information. However great their awareness-raising potential, if making the invisible visible becomes a goal in and of itself then we risk the opportunity to research and design new technologies that allow people to directly act on that knowledge. Put a bit differently, the sentient city may emerge as an informational city but not as an actionable city.

Unique amongst the cases in this volume, JooYoun Paek and David Jimison's Too Smart City project portrays urban street furniture "transformed into representatives of sanitation and control." The notion that designers and architects use materials to shape behavior is not new, and here we may recall Latour's exemplary discussion of speed bumps, or "sleeping policemen," as forms of non-human speed control that drivers regularly negotiate in urban environments. Similarly, architects and urban planners often use special hardware to prohibit skateboarding and sleeping on particular surfaces, or lighting and sound to discourage other unwanted behaviors in public spaces. However, the project anticipates a highly technologized city run amok, where rubbish bins, signs and benches are "rendered non-functional by their overly enthusiastic usage of computational intelligence." Where Natural Fuse raises critical future environmental issues, Too Smart City raises equally critical social and cultural issues. Explicitly positioning the technological objects as failures rather than successes, the project offers an explicit critique of one possible urban future. A different sort of expectation also emerges from this project: that technological research and development need not be serious. Playfully critical approaches to design have been advanced by many practitioners, but Dunne and Raby have also suggested that "Humour is important but often misused. Satire is the goal. But often only parody and pastiche are achieved. These reduce the effectiveness in a number of ways...The viewer should experience a dilemma, is it serious or not? Real or not? For critical design to be successful they need to make up their own mind." Using these criteria, I am not convinced that Too Smart City fully succeeds as satire or critical design. Nonetheless, if more playful and critical approaches were expected and promised I do believe that they offer very interesting alternative or complementary trajectories along which we might proceed.

In summary, and to reiterate my original point, I am interested in how future-oriented expectations begin to do things today. We can already see that current research and design paradigms are creating path dependencies in the areas of techno-urban transmissibility and extensibility. In other words, it is already taken for granted that pervasive urban computing is inevitable, but we have also come to hope, expect and promise that everyday life will be made better as long as we are able to extend and transmit the *right* kinds of urban experience. By gathering projects together in a publication like this, we begin to articulate what the most important or relevant issues are, and we begin to recruit people into particular roles and certain activities that can explore or resolve these issues. If the case studies presented in this volume are successful in creating expectations that truly affect or move people, then their visions for sentient cities will mobilise some research and design activities, but not others. We might see, for example, greater concern over the amount of energy needed to power these dreams, or how designing for safety and convenience can result in unwanted surveillance and control. We might also find ourselves so preoccupied with providing information that we forget to encourage people to act on this knowledge, or consider what might happen if people are unable or unwilling to do so. But I would like to end with the reminder that we can choose to be recruited into each other's visions or we can reject or re-configure these, and other, hopes, expectations and promises. After all, the future still has to be made today.

REFERENCES

— Callon, Michel. (1986) "Some Elements of a Sociology of Translation: Domestication of the Scallops and the Fishermen of St Brieuc Bay." In Power, Action and Belief: A New Sociology of Knowledge, edited by John Law, pp. 196-233. London: Routledge.

— Crang, Mike. (2000) "Urban morphology and the shaping of the transmissable city." City 4(3):303-315.

— Dunne, Anthony and Fiona Raby. (n.d.) "Critical Design FAQ." Available online at http://www.dunneandraby.co.uk/content/bydandr/13/0 (Last viewed 30 September, 2010)

— Latour, Bruno. (1987) Science in Action: How to Follow Scientists and Engineers through Society. Cambridge: Harvard University Press.

— Michael, Mike. (2006) Technoscience and Everyday Life. Maidenhead: Oxford University Press.

POSTSCRIPT: NOTES ON SURVIVAL IN THE SENTIENT CITY

MARK SHEPARD

Science fiction is interesting just as much for the future it invents as for the things in the present it projects forward into that future. The design fictions it imagines are always a hybrid of the novelty of things yet to exist and the durability of some things that somehow continue to survive, at times almost in spite of this future. Take the urban tableau of a future dystopian Los Angeles as projected in Blade Runner, director Ridley Scott's 1982 film based loosely on Philip K. Dick's 1968 novel Do Androids Dream of Electric Sheep? Set in the year 2019, the film tracks the hunt for renegade genetically engineered organic robots called "replicants" through the streets of the city. As the film opens, we find Harrison Ford's character lunching at a noodle stand we might find in present-day Hong Kong. Another scene casts the 1939 Spanish Revival-style Union Station in downtown LA as the LAPD headquarters. The film's climax takes place in George H. Wyman's 1893 Bradbury Building, here repurposed as an apartment building home to the toymaker Sebastian. This intricate portrait of a future saturated with its past invokes an atemporal urban space where the city is characterized by what survives as much as by what evolves.

The Living City Survival Kit, an image by Warren Chalk appearing in the issue of Living Arts that accompanied the Living City exhibition at the ICA in 1963, offers a different urban tableau. Displaying a collection of common, portable, disposable, everyday, mass-produced consumer goods we are asked to consider as essential to city life, the image projects an urban reality that contrasted sharply with certain post-war ideologies of modern architecture and urban planning in the UK, which viewed issues of gender, desire, pop-culture, and consumerism as outside the purview of their fields.[1] Archaeologist Greg Stevenson has referred to an archaeology of the contemporary past as "the design history of the everyday," where common objects drawn from daily life do not simply (passively) reflect cultural forces (trends in taste and fashion, for example) but also actively participate in shaping the evolving social and spatial relations between people and their environment.[2] Read against Jose Luis Sert's 1942 book Can Our Cities Survive?, which outlined the rationalist architecture and urban planning principles of the Congres internationaux d'architecture moderne (CIAM), Chalk's Survival Kit suggested that not only the *vitality* of urban life was itself at risk, but also the architect's role in shaping that environment.

What would an archaeology not of the contemporary past but of the near future — understood as design fictions for tomorrow's cities — reveal about what will survive in the Sentient City? What are the implications for *privacy* in a city where our mobility patterns and transaction histories are stored, aggregated and sold to large corporations? How are conditions of *autonomy* altered when urban infrastructures become responsive to our movements and transactions? In whom, or perhaps what, are we going to need to place our

1. Sadler, Simon. "The Living City Survival Kit: a portrait of the architect as a young man," Art History, Vol. 26 No. 4 (September 2003) pp. 556–575
2. Stevenson, Greg. "Archaeology as the design history of the everyday" in V. Buchli and G. Lucas (eds.) Archaeology of the Contemporary Past, (London: Routledge, 2001) p. 53.

Living City Survival Kit, Warren Chalk + Ron Herron, © Archigram 1963.

trust when accountability is dispersed across an ensemble of networked human and non-human actors? What happens to *serendipity* when browsing the city on foot is replaced by searching it using geo-located datasets? These are just a few of the questions around which a Survival Kit for the Sentient City could form.

CONTRIBUTORS

Tony Bacigalupo is cofounder and Mayor of New Work City, a coworking space and center for independents in Manhattan. Since 2007, Tony has been helping to organize coworking communities and various other local efforts in NYC, including BarCamp, Girl Develop It, Breakout Festival, and Jelly. He was recently featured on the cover of October 2010's Inc Magazine. He co-authored "I'm Outta Here: How coworking is making the office obsolete" and has spoken at SXSW, MakerFaire, NY Tech Meetup, and Influencer10. Tony's work continues to focus on helping people develop better careers as independent professionals. His latest effort, NWCU, aims to build a people-powered school where existing and aspiring entrepreneurs and freelancers gather to share knowledge and help one another.

David Benjamin along with Soo-in Yang, created The Living in 2004 and launched the Living Architecture Lab in 2005. The practice and the lab emphasize open source research and design, and work is considered to be part of a broad eco-system of design—each project is a loop of research connected to other loops developed before and after it. Recent projects have explored architecture in the context of public space, ubiquitous computing, and the environment. They involve dynamic building facades, constructions in highly-visible urban areas, interfaces to environmental quality, platforms for collective discussion and action, and new forms of text-message communication with architecture.

Keller Easterling is an architect, urbanist, and writer who researches infrastructure, networks and the political implications of certain contemporary spatial conditions. Her latest book, Enduring Innocence: Global Architecture and Its Political Masquerades (MIT, 2005), researches familiar spatial products that have landed in difficult or hyperbolic political situations around the world. Her previous book, Organization Space: Landscapes, Highways and Houses in America, applies network theory to a discussion of American infrastructure and development formats. A forthcoming book, Extrastatecraft, examines infrastructure as a medium of polity. Her work has been widely published in journals such as Grey Room, Volume, Cabinet, Assemblage, Log, Praxis, Harvard Design Magazine, Perspecta, Metalocus, and ANY. Easterling is an Associate Professor at the Yale School of Architecture.

Laura Forlano is a Postdoctoral Associate in the Interaction Design Lab in the Department of Communication at Cornell University. Forlano received her Ph.D. in Communications from Columbia University in 2008. Her dissertation, "When Code Meets Place: Collaboration and Innovation at WiFi Hotspots," explores the intersection between organizations, technology (in particular, mobile and wireless technology) and the role of place in communication, collaboration and innovation. Forlano is an Adjunct Faculty member in the Design and Management department at Parsons and the Graduate Programs in International Affairs and Media Studies at The New School where she teaches courses on Innovation, New Media and Global Affairs, and Technology and the City. She serves as a board member of NYCwireless and the New York City Computer Human Interaction Association.

Matthew Fuller is author of various books including Media Ecologies, materialist energies in art and technoculture, Behind the Blip, essays on the culture of software and the forthcoming Elephant & Castle. With Usman Haque, he is co-author of a pamphlet "Urban Versioning System v1.0" in Situated Technologies Pamphlets Series, and with Andrew Goffey, co-author of the forthcoming Evil Media. Editor of Software Studies, a lexicon, and co-editor of the new Software Studies series from MIT Press, he is involved in a number of projects in art, media and software and works at the Centre for Cultural Studies, Goldsmiths, University of London.

Anne Galloway brings a strong background in sociology and anthropology to her position as Senior Lecturer, Design Research in the School of Design, Victoria University of Wellington, New Zealand. Her research explores the roles that cultural studies and design research can play in supporting public understandings of new technologies and promoting more active participation in their development and implementation.

Usman Haque is director Haque Design + Research Ltd, founder of Pachube.com and CEO of Connected Environments Ltd. He has created responsive environments, interactive installations, digital interface devices and mass-participation performances. His skills include the design and engineering of both physical spaces and the software and systems that bring them to life. He has been an invited researcher at the Interaction Design Institute Ivrea, Italy, artist-in-residence at the International Academy of Media Arts and Sciences, Japan and has also worked in USA, UK and Malaysia. As well as directing the work of Haque Design + Research he was until 2005 a teacher in the Interactive Architecture Workshop at the Bartlett School of Architecture, London.

Dan Hill is a designer and urbanist whose work explores how real-time information networks change streets and cities, neighborhoods and organizations, mobility and work, play and public space. His writing on urban informatics has been published in books, journals and periodicals internationally. Based in Sydney, Australia, Hill is Leader of Foresight + Innovation for the Australasian region of global multidisciplinary design firm Arup, and their global leader for urban informatics.

Natalie Jeremijenko, director of the xDesign Environmental Health Clinic at NYU, is an artist whose background includes studies in biochemistry, physics, neuroscience and precision engineering. Jeremijenko's projects explore socio-technical change, and have been exhibited by several museums and galleries, including the MASSMoCA, the Whitney, Smithsonian Cooper-Hewitt. A 1999 Rockefeller Fellow, she was recently named one of the 40 most influential designers by I.D. Magazine. She is an Associate Professor of Visual Art at NYU, a visiting professor at Royal College of Art in London, and an artist not-in-residence at the Institute for the Future in Palo Alto. Previously, Jeremijenko was a member of the faculty in the Visual Arts at UCSD and in Engineering at Yale.

David Jimison is an artist, inventor, & student focused on the intersections of pervasive technology (eg. mobile, augmented reality, embedded computing) and creative cultures (eg. art, music, fashion.) David's consulting work and commercial inventions are done through his company Fever Creative. As an artist, David creates immersive interactive installations that elicit new behaviors, and thereby experiences, to the people embedded within them, engaging them in momentary, festive, and outlandish landscapes.

Omar Khan is an architect and educator whose research explores the role of embedded computational technologies for designing responsive architecture. His research focuses on augmenting environments with sensing and actuating technologies, rethinking material substrates and assemblies, and theorizing ways to develop mutualist relationships with our built environment. He is co-author with Philip Beesley of "Responsive Architecture, Performing Instruments" and is a co-editor of the Situated Technologies Pamphlets Series, published by the Architectural League of New York. He is an Associate Professor of Architecture at the University at Buffalo, where he co-directs the Center for Architecture and Situated Technologies.

MIT's SENSEable City Lab has become a leader in creatively investigating and analyzing the interface between people, technology and the city. SENSEable's work draws on diverse fields such as urban planning, architecture, design, engineering, computer science and social science to capture the multi-disciplinary nature of urban problems. SENSEable's approach begins with a vision for an "urban demo" tailored to a particular city's needs, and is motivated by unique challenges or opportunities within the city due to advances in digital technologies. These urban demos are showcased at large public venues to stimulate debate between citizens, public administrators and industry representatives. Since 2004, the Lab's activities have grown rapidly, with 63 researchers completing 35 projects from around the world. The Lab has produced over 200 scientific publications and its work has been covered over 1,000 times by global media and press outlets like The New York Times, The Economist, National Geographic, CNN and Wired among others.

JouYoun Paek is an artist and interaction designer born in Seoul and based in New York. She has created interactive objects and installations that reflect on human behavior, technology and social change. Her art has been displayed by the Museum of Modern Art New York, Postmasters Gallery, EYEBEAM, Museum of Science Boston and Seoul Museum of Art.

Saskia Sassen's work focuses on the social, economic and political dimensions of globalization, immigration, global cities (including cities and terrorism), new networked technologies, and changes within the liberal state that result from current transnational conditions. Her books include The Global City (Princeton University Press, 1991; New edition 2001) and Territory, Authority, Rights: From Medieval to Global Assemblages (Princeton University Press 2008). She is the Robert S. Lynd Professor of Sociology and Co-Director of the Committee on Global Thought, Columbia University.

Trebor Scholz is a writer, conference organizer, Assistant Professor in Media & Culture, and Director of the conference series The Politics of Digital Culture at The New School in NYC. He also founded the Institute for Distributed Creativity that is known for its online discussions of critical Internet culture. Trebor is co-editor The Art of Free Cooperation and editor of The Internet as Playground and Factory, forthcoming from Routledge. Forthcoming edited collections include The Digital Media Pedagogy Reader and The Future University, both by iDC. His recent writing examines the history of digital media activism, the politics of Facebook, limits to accessing knowledge in the United States, and mobile digital labor. His forthcoming monograph offers a history of the Social Web and its Orwellian economies. He holds a PhD in Media Theory and a grant from the John D. & Catherine T. MacArthur Foundation.

Mark Shepard an artist, architect and researcher whose work addresses new social spaces and signifying structures of contemporary network cultures. He is an editor of the Situated Technologies Pamphlets Series, published by the Architectural League of New York and co-author of "Urban Computing and its Discontents" with Adam Greenfield. Other publications include "Tactical Sound Garden [TSG] Toolkit", in 306090 v.9–Regarding Public Space, and "Extreme Informatics: Toward the De-saturated City" in Urban Informatics: The Practice and Promise of the Real-time City. Mark is an Assistant Professor of Architecture and Media Study at the University at Buffalo, where he co-directs the Center for Architecture and Situated Technologies.

Antonina Simeti is a strategic consultant at DEGW. She works with universities, museums and corporations to leverage building and workplace design to improve organizational performance. She has a special interest in knowledge sectors and the creative economy, and the spaces in which innovation happens. At DEGW she is currently leading a research program on new practices in workplace strategy (including design, technology and human resources solutions), building a metrics framework to measure the impact of workplace design on productivity at a global financial services corporation, and developing the building program for a major contemporary art museum expansion. Antonina earned her Masters in City Planning from MIT and her B.A. in economics from Tufts University.

Hadas Steiner is an architectural historian whose research concentrates on the cross-pollinations of technological and cultural aspects of architectural fabrication in the postwar period. She is the author of Beyond Archigram: The Technology of Circulation (Routledge 2007). Work in progress includes manuscripts on the photographic documentation conducted by Reyner Banham while in Buffalo, the techno-zoological architecture of Cedric Price, as well as on the architecture of extreme conditions, including the work of John McHale. Hadas is an Associate Professor of Architecture at the University at Buffalo, State University of New York.

Anthony Townsend is Research Director at the Institute for the Future, an independent research organization based in Silicon Valley, California. His research and consulting focus on several inter-related issues: mobility and urbanization, innovation systems and innovation strategy, science and technology parks and economic development, and sustainability and telework. He was one of the original founders of NYCwireless, a pioneer in the municipal wireless movement that promotes the use of public-access Wi-Fi in the development of local communities. He holds a Ph.D. in urban and regional planning from MIT. He lives in Hoboken, New Jersey, the birthplace of baseball and Frank Sinatra.

Soo-in Yang, along with David Benjamin, created The Living in 2004 and launched the Living Architecture Lab in 2005. The practice and the lab emphasize open source research and design, and work is considered to be part of a broad ecosystem of design—each project is a loop of research connected to other loops developed before and after it. Recent projects have explored architecture in the context of public space, ubiquitous computing, and the environment. They involve dynamic building facades, constructions in highly-visible urban areas, interfaces to environmental quality, platforms for collective discussion and action, and new forms of text-message communication with architecture.

Martijn de Waal is a writer and researcher based in Amsterdam whose work focuses on digital media and public culture. His writing on urban computing and locative media has been published in books, journals and periodicals internationally. With Michiel de Lange he founded The Mobile City, an initiative focusing on digital media and urban culture. He is a PhD candidate at the University of Groningen in the New Media, Public Sphere and Urban Culture research group.

Kazys Varnelis is an architectural historian whose writing on urban infrastructure and contemporary network culture is widely read. He directs the Network Architecture Lab at the Columbia University Graduate School of Architecture, Planning, and Preservation. He is editor of the Infrastructural City. Networked Ecologies in Los Angeles, Networked Publics and The Philip Johnson Tapes: Interviews with Robert A. M. Stern, all published in 2008. Varnelis is a co-founder of the conceptual architecture/media group AUDC, which published Blue Monday: Absurd Realities and Natural Histories in 2007 and has exhibited widely in places such as High Desert Test Sites. He has also worked with the Center for Land Use Interpretation, for which he produced the pamphlet Points of Interest in the Owens Valley.

Gregory Wessner is Digital Programs and Exhibitions Director for the Architectural League. Recent exhibitions curated by Wessner include The City We Imagined/The City We Made: New New York 2001-2010, New New York: Fast Forward, and 13:100 | Thirteen New York Architects Design for Ordos. Wessner has produced more than two hundred podcasts and also edited publications including The Architectural League: 125 Years and Travel Reports from the Deborah J. Norden Fund. He formerly served as Chief Administrator of the National Academy School of Fine Arts and White Columns.

Mimi Zeiger founded loud paper, an architecture zine and now blog, in 1997. The publication received grant awards from the Graham Foundation for Advanced Studies in the Fine Arts and the LEF Foundation. Zeiger is author of New Museums: Contemporary Museum Architecture Around the World, Tiny Houses and Micro Green. A Brooklyn-based freelancer, she writes on art, architecture, and design for a variety of publications including The New York Times, Metropolis, Dwell, Azure, and Architect, where she is a contributing editor. She has taught at Parsons, the California College of the Arts (CCA) and at the Southern California Institute of Architecture (SCI-Arc.) Her cross-disciplinary seminars explore the relationships between architecture, art, urban space, and popular culture. She holds a Master of Architecture degree from SCI-Arc and a Bachelor of Architecture degree from Cornell University.

http://survival.sentientcity.net, Mark Shepard, 2010.